Blockchain-based Cyber Security
Applications and Paradigms

Editor

Kaushal Shah

Department of Computer Science and Engineering
Pandit Deendayal Energy University, Gandhinagar, India

CRC Press is an imprint of the
Taylor & Francis Group, an **informa** business
A SCIENCE PUBLISHERS BOOK

First edition published 2024
by CRC Press
2385 NW Executive Center Drive, Suite 320, Boca Raton FL 33431

and by CRC Press
4 Park Square, Milton Park, Abingdon, Oxon, OX14 4RN

CRC Press is an imprint of Taylor & Francis Group, LLC

Library of Congress Cataloging-in-Publication Data (applied for)

ISBN: 978-1-032-48543-0 (hbk)
ISBN: 978-1-032-48544-7 (pbk)
ISBN: 978-1-003-38957-6 (ebk)

DOI: 10.1201/9781003389576

Typeset in Times New Roman
by Prime Publishing Services

Preface

This book summarizes the key areas of cyber security. The book focuses on the specific paradigm of blockchain technology to enforce cyber security. The book demonstrates the deployment of various security attacks in a real-time network environment. The challenges related to cyber security and the solutions based on Software Defined Networks are discussed in detail in the book. The book uniquely presents the solutions to cyber security attacks by considering various real-time applications based on the Internet of Things, Wireless Sensor Networks, Cyber-Physical Systems, and Smart Grids. The book is helpful for academicians and research scholars worldwide working in the cyber security field. The book is also beneficial to industry experts working in the cyber security field. Undergraduate students who are exploring the area of cyber security can also utilize the contents of the book. The book can help design an advanced-level course on cyber security for undergraduate and postgraduate students. It is also helpful for industry penetration testers and vulnerability assessment experts. The book emphasizes the following significant aspects in the field of cyber security in total depth: Application Layer Attacks and Detections, Security & Privacy Issues in Cloud Computing, Zero-Day Attack Detection, Security Analysis in Smart Environments, Security and Privacy Issues in Searchable Encryption in Cloud, Provable Security, Security of IoT platforms, Volatile Memory Forensics, Formal methods and verification for security, Security Aspects of Blockchain Technology, Challenges in Cyber Security-based Internet of Things, Realizing the Cloud-based Blockchain Platform, SDN enabled Blockchain network.

Contents

Preface iii

1. **A Survey of Blockchain for the Mitigation of DDoS Attacks in IoT Networks** 1
Umamaheswari Rajasekaran, A. Malini and *Vandana Sharma*

2. **Blockchain-Enabled SDN in Resource Constrained Devices** 12
Anil Carie, M. Krishna Siva Prasad and *Satish Anamalamudi*

3. **Computer Vision Mapping for Parking Space Counter** 26
Atul Srivastava and *Abdul Azeem Khan*

4. **Innovative, Conceptual, and Analytical Approaches for Cyber Security, IoT and Blockchain** 50
Ramesh Ram Naik, Umesh Bodkhe, Rohit Pachlor, Sunil Gautum and *Sanjay Patel*

5. **Mitigating Security Vulnerabilities in the Internet of Things: An Examination of Blockchain-based Solutions** 66
Ansh Suresh Bhimani, Rohan Rakeshkumar Shah and *Kaushal Arvindbhai Shah*

6. **A Comprehensive Assessment of Modern Blockchain-based Secured Health Monitoring Systems** 99
Swati Manekar and *Umesh Bodkhe*

7. Security of IoT Platforms: Current Challenges and Future Directions **113**
Rashi Sahay and *Shanu Khare*

8. Security Aspects of Blockchain Technology **128**
Ramu Kuchipudi, T. Satyanarayana Murthy, Ramesh Babu Palamakula and *R.M. Krishna Sureddi*

Index **141**

Chapter 1

A Survey of Blockchain for the Mitigation of DDoS Attacks in IoT Networks

Umamaheswari Rajasekaran,[1] *A. Malini*[1,2,*] and *Vandana Sharma*[3]

1. Introduction

Blockchain is a public distributed ledger technology that is used for recording transactions, thus encouraging tamper-free, immutable, non-deniable, secure, decentralised transactions in the network. Conventional Host based addressing systems are susceptible to Single Point of Failure. In Information Security, Hashing is a technique used for ensuring the integrity of the data stored. But, for example, if the database is hacked and the digest is tampered or the records are deleted, it is

[1] Computer Science and Business Systems, Thiagarajar College of Engineering, Madurai, India.

[2] ORCID ID:0000-0002-3324-5317.

[3] Amity Institute of Information Technology, Amity University, Noida, India.

Emails: umatwothousandandthree@gmail.com; vandana.juyal@gmail.com

* Corresponding author: amcse@tce.edu

impossible to recover the original data without a backup. In the case of Blockchain, the copy of data is stored in every participatory node of the network. Thus, even if a node is hacked, no availability of the data is still ensured. The core concepts of blockchain such as Mining, Consensus Algorithms, Merkle Tree, Hashing Algorithms, and Encryption ensure the Confidentiality, Integrity, and Availability (CIA) of data.

Theoretically, Blockchain is a chain of blocks, where each block stores traces of the unanimously stored transactional data. The process of adding new blocks to the blockchain is called Mining and the nodes adding new blocks are called Miner Nodes. Transaction pool is a collection containing unconfirmed transactions. In simple words, it consists of transactions that are yet to be mined. There is a paramount distinction between Transaction Validation and Transaction Mining. Mining is the process of adding the blocks to the blockchain, whereas Validation is the process of verifying if the transactions conform to certain protocols within the network. Whenever the request for the transaction has been received, it has to be validated. Validator nodes are the same as the miner nodes; thus, it promotes security in the network. Hashing is the process of scrambling the message into an unidentifiable format called the digest, such that any small change in the original message can introduce huge differences in the recomputed hash for the altered message. Hash functions are irreversible. This means that it is possible to generate digest from the message using a suitable algorithm; however, it is impossible to recompute the original message from the digest. The property of determinism of the hash functions states that the hash function should compute the same digest for the same message given as input several times.

Merkle Tree (Fig. 1) is a data structure that reduces the overhead to store the copies of actual transactions within a block in the network. Every block records a huge number of transactions. Initially, the transactions are paired up and the hash of the pairs are computed. In the case of an odd number of blocks, the last block is replicated. The computed hash values are again paired and hashed. This process is repeated until the branches reduce to a single node.

Every block header consists of previous blocks' hash pointer, Merkle Root, Block Hash, and Nonce. Hash pointer stores the hash of the previous block and points to the previous block. Thus, any change made in the previous block, can be easily detected as the hash value stored in the proceeding block will not match with the hash of the value in the

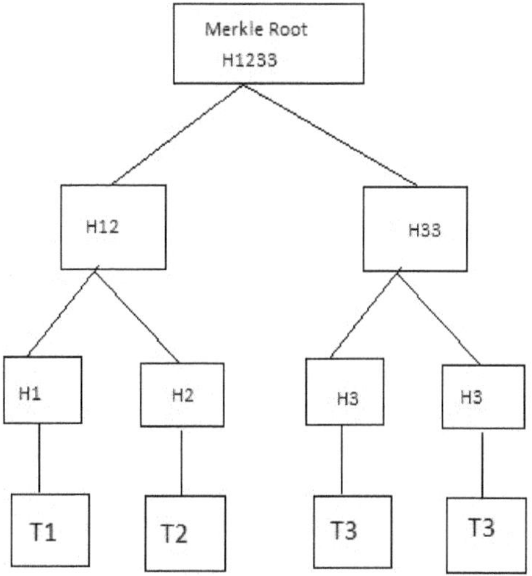

Fig. 1. Merkle Tree representation (Block with odd number of transactions).

previous block. SHA-256 that produces a 256-bit output is the popularly used hashing algorithm in Bitcoin. A byte is made up of 8 bits. 256 bits constitute 32 bytes. These 256 bits are grouped into 64 characters of 4 bits – hexadecimal representation producing 64 hexadecimal characters as output. This is represented in (Fig. 2). Figure 3 shows the SHA-256 output for a slightly changed Input message 'ABCd' other than 'ABCD'. Comparing the Hex Output in Figs. 2 and 3 we can see that for a slight change in input, there is a drastic change in the digest produced. This behaviour of hashing functions is called as Avalanche effect. This effect thus ensures absolute randomness and makes it difficult to guess the input message using the digest by techniques of statistical modelling.

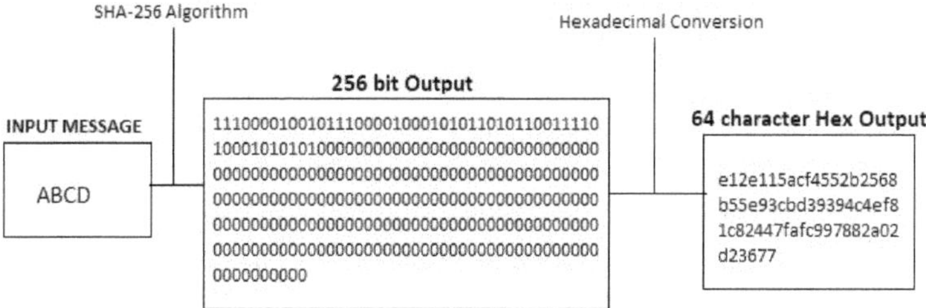

Fig. 2. SHA-256 algorithm output representation.

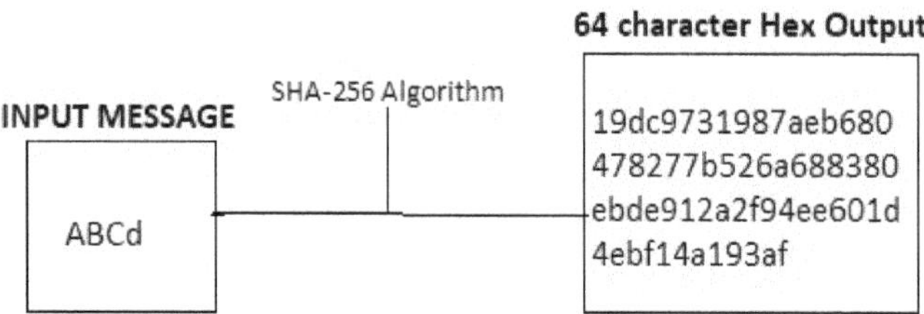

Fig. 3. SHA-256 output for slightly changed data.

Dictionary attack is the most common way to crack a hashing algorithm, however it takes ample time. It is very difficult to crack the hashing algorithm as the digest is extremely different from the input.

Consensus algorithms are used to ensure the order of writing blocks into the blockchain. It authorises the nodes to become miner nodes, thus promoting security and consistency among the nodes to hold the same instance of the block. Proof of Work (PoW), Proof of Stake (PoS), Proof of Elapsed Time, Proof of Capacity, and Proof of Authority are some of the commonly known consensus algorithms. Unconfirmed transactions are selected from the pool, based on consensus they are added to the longest chain of the block.

Proof of Work: This consensus algorithm expects the nodes to solve a complex mathematical puzzle that has huge computational requirements to arrive at the answer. The node that finds the solution first becomes the miner node and adds the block to the chain. The miner node is rewarded in crypto units. The complexity of the problem increases with increase in computational power. One of the commented drawbacks of the PoW consensus is that it does not support recording instantaneous transactions, rather it takes a few minutes to mine the block in the network. Also, there is more probability for the node having higher computational power to become the miner more frequently.

Proof of Stake: Unlike the PoW, the PoS is more centralised towards the deposits of the stake coins in the network. The chance to become the next node adding a transaction in the block is directly proportional to the stake coin deposit. They are penalized for fraudulent transactions and rewarded as a fee for recording true transactions. Even though this emphasizes centralisation towards the stake coins, it is advantageous in comparison with the PoW consensus as it does not consume much energy.

Proof of Elapsed Time: It is a consensus algorithm released by Intel in 2017. Every participating node generates a random number. Every node waits for the time specified by the generated number. The node which generated the smallest random number becomes the winner. The generation of the random number is controlled by the system (SGX). This ensures fairness in the network as every node has an equal probability to become the next winner.

Proof of Authority: Unlike PoW and similar to PoS, it does not involve solving puzzles to attain the consensus within the network. Unlike PoS, validators don't stake coins but their own reputation in the network. It works by choosing a limited number of validators and is used extensively for private blockchains and has a higher transaction rate than PoS and PoW. However, as there are only a limited number of trusted validator nodes in the network, it is more centralized.

Sybil attacks are common types of attacks encountered in a blockchain network. When one node pretends to be many or when many unique identities are controlled by one identity, it is called a sybil attack [1]. The 51% attack is more popular in blockchain networks. Hash rate is a measure of computational power spent by the blockchain nodes collectively in unit time. In PoW consensus, the number of guesses made per second to solve the puzzle determines the hash rate of the network. By the 51% attack, the attacker controls more than 50% hash rate of the network. Thus, there is a higher chance of data tampering, double spending, changing the order of transactions, etc. However, controlling more than 50% of the network is practically impossible in large distributed blockchain networks. But, history in the past years has recorded the occurrence of 51% attack on small power blockchain networks like Bitcoin Gold. In 2018, Bitcoin Gold was hit by 51% attack which led to 18 million dollars of BTG [2]. Similarly, Litecoin cash, Hanacoin also faced 51% attacks [3].

This chapter is organised with Section 2 elaborating the potential use cases of Blockchain for preventing cyber-attacks especially DDoS (distributed denial of service) in the IoT networks. Section 2 also includes a detailed discussion of False Data Injection attacks in Smart Grids. Section 3 comprises Conclusion.

2. Literature Survey

A. Blockchain in the Mitigation of DDoS Attacks in IoT Networks

Minoli and Occhiogrosso [4], presented a survey discussing the scopes of Blockchain technology to solve security issues in different IoT environments. The point where the data are collectively stored or processed to extract information is considered as the active focus point for cyber-attacks. Blockchain is considered to be an effective and promising technology for addressing the CIA (central intelligence agency) issues in the conventional IoT (internet of things) networks that are both memory constrained and power constrained in general. Even though the constraints of resources in the IoT network makes it difficult to integrate with Blockchain globally, it has been recommended for specific established networks such as e-healthcare, smart grids, etc. Singh et al. [5] analysed the potential of Blockchain Technology for preventing DDoS attacks in IoT networks. Ethereum-based blockchains are mainly designed for executing smart contracts which brings verified and decentralised automation within the networks. Smart contracts based DDoS attack mitigation strategy in Ethereum blockchain attracted the attention of the researchers. It has been concluded that the presence of scalability issues in the blockchain makes it difficult for large-scale implementation. Abou et al. [6] proposed a blockchain-based DDoS attack mitigation called Cochain-SC that makes use of smart contracts and SDN (software-defined networking). They proposed a 3-scheme-based Intra domain attack prevention using SDN. The autonomous systems in the network have been grouped into 3: source, intermediate, and the destination. The origin of the attack (attacker) is the source domain. The systems that forward the traffic are classified as the intermediate domains. The target system of the attack is the destination domain. Smart Contract-based DDoS attack mitigation strategy has been designed where the collaborators in the network share the copies of the malicious IP addresses. It has been concluded that the designed method effectively arrests the attack nearer to the source as the other domains within the network are informed with the trace of malicious IP addresses. Jia and Liang [7] proposed a DDoS attack detection framework based on parallel intelligence and ACP theory. In the research work, the true blockchain has been coupled to interact with the simulated blockchain using the ACP theory in parallel. Detection in the simulated blockchain helps to defend the attacks in the true blockchain. AdaBoost and Random Forest algorithms have been used in the first and second blockchains. The experimental results have been concluded based on the KDD Cup 99 dataset. Manikumar and Maheswari [8] proposed a blockchain-based decentralised ledger that tracks the malicious IP address detected by the trained classifiers in the network. K-NN, Random Forest, and Decision Tree classifiers have been implemented. hping3 command in linux has been used to simulate a DDoS attack. A certain threshold based on the Time Stamp has been set to manage blocking and unblocking of the malicious IP addresses. The models have been trained using the CICDDoS2019 dataset, and it has been concluded that the Random Forest algorithm provided a better accuracy of 95%. Yeh et al. [9] presented a smart-contract based decentralised data sharing platform (consortium blockchain) for sharing the DDoS attack information such that the availability of the attack information for Security Operation Centre increased. A verification mechanism and Bloom filter to ensure privacy has been enforced. Liu and Yin [10] initiated an LSTM-CGAN architecture for generating samples of LDDoS-attacks in blockchain-based wireless networks. It has been concluded that the generated samples well mimicked the public and private datasets, and is suitable for training and real-time deployment of models for detecting LDDoS attacks in blockchain-based wireless networks.

B. Smart Grids and False Data Injection Attacks

Energy demands rise exponentially with the increasing population. Need for automation and centralised management of both the production and the distribution of power has resulted in the design of smart grids that are cables of electricity distribution with automation that also involves collection, storage, and processing of data for increasing efficiency. Unlike the conventional energy distribution units, smart grids have both power transmission lines and data lines. National Institute of Standards and Technologies (NIST) defines the structure of smart grids to be composed of generation, transmission, distribution, operation, electricity markets, service providers, and customer domains. With the increasing penetration of Renewable sources of energy into the nation's power generation framework, the adoption of smart grids to manage power distribution lanes across different sectors also increases. Unlike the conventional sources of power generation, Renewable Sources of energy show stochasticity in the measure of the energy output due to the variations in the environmental factors, like cloudiness influencing the solar power generation, formation of cyclones influencing wind energy, etc. The poor design of resources in terms of security used in the design of smart grid increases the vulnerability to both physical attacks and cyber-attacks [11]. Data related to crucial power systems of the industry or nation require both confidentiality, integrity, and availability ensuring security (CIA standards). Breaking the confidentiality of the data can have significant impacts on the economic level, as the true architecture, plan, and the future state of power demands can be easily predicted. Stuxnet is one of the most famous worm-based cyber-attack, where the goal was to damage the physical components of the Iranian nuclear power plants itself. The true target of the Stuxnet was the controllers that contribute to the mechanics of the physical system. Stuxnet attack has been described as a stand-alone that didn't require an internet connection. As soon as the worm was infected to the Windows OS-based PC, Stuxnet searched for a specific model type controller from Siemens manufacturer [12]. Reports claim that, in 2015, Black Energy malware caused an unprecedented power outage for around 6 hours [13]. Supervisory Control and Data Acquisition Systems (SCADA) are types of Industrial control systems, a combination of both hardware and software that forms the core design of Smart Grids. Its design and implementation is based on the network of ICT. Ferrag et al. [14] present fog-based SCADA architecture for Smart Grids, and classify the solutions into four categories: Authentication, Privacy-preserving, Key Management, and Intrusion Detection Systems (IDS). The paper also provides a nine categorization for the types of IDS. It has been concluded that significant research to detect bad data injection, blockchain-based solutions are yet to be focussed in the literature. Gunduz and Das [15] classified the attacks and the attacker types in the smart grid. It has been reported that there are chances for both conscious and unconscious cyber-attacks by the customers, employees involved in the network. Attacks that compromise the data integrity from the customer side may be done for fiscal benefits. Availability of the data within the network must be ensured for maintaining the secure operating state of the network. The paper also highlights nine security requirements in smart grids.

False Data Injection (FDI) is one of the most widely studied attacks in relation to the smart grid environment. Any change in the original data present in the database, adding new fake values, deleting values from the database are different ways by which FDI attacks can be instigated by attackers. This attack that compromises data integrity can have critical impacts fiscally. This poisoned data can also have serious impacts in the model behaviour if used for training Energy forecasting models without proper quality checks [16]. Musleh et al. [17] presented a detailed survey of cyber-attacks in smart grids. FDI Attacks (FDIA) has been categorised into four: Physical-layer based FDIA, Network-based FDIA, Communication-based FDIA, Cyber based FDIA. Among the Supervised Algorithms, SVM has been reported as the most used algorithm for FDI detection. Increasing the number of hidden layers in the neural network has been observed to increase the extent of FDI detection. Computational Complexity involved in training the data-driven approaches to FDI detection has been discussed as an issue. Design of FDI systems with reduced false alarms, Detection Speed, etc. have been concluded as the yet-to-be met objectives in the design of FDI systems for smart grid applications. Esmalifalak et al. [18] proposed an SVM-based (support vector machine) FDI detection where PCA (prompt corrective action) has been used to reduce the dimensionality of the data. It has been claimed that the proposed method was able to separate the actual data from the tampered data as the true data was governed by physics whereas the tampered data has been observed to have randomness. Lu et al. [19] highlighted the data security issues in the edge computing architecture and proposed a blockchain-based data aggregation framework for edge computing of Smart Grid Data. They considered two types of tampering in their system model. As discussed, the attacker may either eavesdrop data packets in the communication channel and tamper it or the attacker can instigate false data injection in the database servers. Both the types have been considered in the User Layer and the Edge Layer of the system model. Paillier Encryption Algorithm has been used. It has been proved that the design ensures confidentiality, integrity, privacy of the users, Authentication, and shows resistance to attacks with reduced computational cost.

3. Conclusion

This study presented a detailed survey of applications of Blockchain technology for preventing cyber-attacks in the IoT network. The widely known distributed technology, with its core properties, finds use cases for preventing cyber-attacks. The core idea of design of a decentralised data transaction platform to exchange malicious IP addresses for mitigating DDoS attacks has been emphasised in many literatures. The commonly known characteristics of the Blockchain network are immutability, decentralization, non-repudiation, and security. The identity of the individual blockchain users is ever held anonymous. Mining pool refers to the collection of resources that share the computing power to mine the next block. The choice of the consensus algorithms defines the energy usage in the network. Proof of Work, Proof of Stake, Proof of Authority, Proof of Elapsed Time, and Delegated Proof of Stake are some of the commonly used Consensus Algorithms. Bitcoin uses the PoW consensus whereas Ethereum currently uses PoS. Based on the application, the consensus algorithms are chosen as every algorithm has both advantages and disadvantages.

This study also surveyed the mitigation approaches for the prevention of False Data Injection attacks in Smart Grids. The keyword search in Google Scholar, '(Cyber Attacks) AND Blockchain', gave around 63,900 results in 0.1 second whereas the keyword search, 'FDI AND Blockchain AND (smart grid)', gave around 1,750 results in 0.14 second. It can be concluded that the application of Blockchain technology to prevent cyber-attacks in Smart Grids is not as actively studied as the research on the scopes of Blockchain technology for preventing cyber-attacks in IoT networks.

References

[1] Douceur, J.R. 2002. The sybil attack. pp. 251–260. *In*: *Peer-to-Peer Systems: First International Workshop, IPTPS 2002 Cambridge,* MA, USA, March 7–8, 2002 Revised Papers 1. Springer Berlin Heidelberg.

[2] https://en.wikipedia.org/wiki/Bitcoin_Gold. [Accessed on Jan. 8th 2023.]

[3] Murtuza Merchant. 2022, Nov., What is a 51% attack and how to detect it. https://cointelegraph.com/news/what-is-a-51-attack-and-how-to-detect-it.

[4] Minoli, D. and Occhiogrosso, B. 2018. Blockchain mechanisms for IoT security. *Internet of Things*, 1: 1–13.

[5] Singh, R., Tanwar, S. and Sharma, T.P. 2020. Utilization of blockchain for mitigating the distributed denial of service attacks. *Security and Privacy*, 3(3): e96.

[6] Abou El Houda, Z., Hafid, A.S. and Khoukhi, L. 2019. Cochain-SC: An intra-and inter-domain DDoS mitigation scheme based on blockchain using SDN and smart contract. *IEEE Access*, 7: 98893–98907.

[7] Jia, B. and Liang, Y. 2020. Anti-D chain: A lightweight DDoS attack detection scheme based on heterogeneous ensemble learning in blockchain. *China Communications*, 17(9): 11–24.

[8] Manikumar, D.V.V.S. and Maheswari, B.U. 2020, July. Blockchain-based DDoS mitigation using machine learning techniques. pp. 794–800. *In*: *2020 Second International Conference on Inventive Research in Computing Applications (ICIRCA)*. IEEE.

[9] Yeh, L.Y., Lu, P.J., Huang, S.H. and Huang, J.L. 2020. SOChain: A privacy-preserving DDoS data exchange service over soc consortium blockchain. *IEEE Transactions on Engineering Management*, 67(4): 1487–1500.

[10] Liu, Z. and Yin, X. 2021. LSTM-CGAN: Towards generating low-rate DDOS adversarial samples for blockchain-based wireless network detection models. *IEEE Access*, 9: 22616–22625.

[11] He, H. and Yan, J. 2016. Cyber-physical attacks and defences in the smart grid: A survey. *IET Cyber-Physical Systems: Theory & Applications*, 1(1): 13–27.

[12] Langner, R. 2011. Stuxnet: Dissecting a cyberwarfare weapon. *IEEE Security & Privacy*, 9(3): 49–51.

[13] Christina Miller. 2021, Nov. Throwback Attack: BlackEnergy attacks the Ukrainian power grid. https://www.industrialcybersecuritypulse.com/threats-vulnerabilities/throwback-attack-blackenergy-attacks-the-ukrainian-power-grid/.

[14] Ferrag, M.A., Babaghayou, M. and Yazici, M.A. 2020. Cyber security for fog-based smart grid SCADA systems: Solutions and challenges. *Journal of Information Security and Applications*, 52: 102500.

[15] Gunduz, M.Z. and Das, R. 2018, September. Analysis of cyber-attacks on smart grid applications. pp. 1–5. *In*: *2018 International Conference on Artificial Intelligence and Data Processing (IDAP)*. IEEE.

[16] Ahmed, M. and Pathan, A.S.K. 2020. False data injection attack (FDIA): An overview and new metrics for fair evaluation of its countermeasure. *Complex Adaptive Systems Modeling*, 8(1): 1–14.

[17] Musleh, A.S., Chen, G. and Dong, Z.Y. 2019. A survey on the detection algorithms for false data injection attacks in smart grids. *IEEE Transactions on Smart Grid*, 11(3): 2218–2234.

[18] Esmalifalak, M., Liu, L., Nguyen, N., Zheng, R. and Han, Z. 2014. Detecting stealthy false data injection using machine learning in smart grid. *IEEE Systems Journal*, 11(3): 1644–1652.

[19] Lu, W., Ren, Z., Xu, J. and Chen, S. 2021. Edge blockchain assisted lightweight privacy-preserving data aggregation for smart grid. *IEEE Transactions on Network and Service Management*, 18(2): 1246–1259.

Chapter 2

Blockchain-Enabled SDN in Resource Constrained Devices

Anil Carie, M. Krishna Siva Prasad* and *Satish Anamalamudi*

1. Introduction

Software-defined networking (SDN) with blockchain integration is an innovative technology that merges the advantages of both SDN and blockchain to deliver a reliable and effective network structure. By introducing a decentralized and immutable ledger, blockchain technology guarantees the safety and accountability of network activities, documenting every transaction and interaction within the network. SDN provides a programmable and flexible network infrastructure that can be dynamically controlled and managed. The combination of these two technologies can offer several advantages, such as improved security, privacy, and transparency, as well as enhanced scalability, reliability, and efficiency. For instance, blockchain-enabled SDN can prevent malicious attacks and ensure the authenticity of network transactions,

Department of Computer Science and Engineering, School of Engineering and Sciences, SRM University AP, Amaravathi, India.
Emails: krishnasivaprasad536@gmail.com; satishnaidu80@gmail.com
* Corresponding author: carieanil@gmail.com

while also providing a high level of privacy protection and traceability. Furthermore, it can support decentralized management and control of network resources, which can increase the network's efficiency and reduce operational costs. In general, the utilisation of blockchain in SDN architecture offers a potential resolution for establishing a secure and effective network infrastructure in the forthcoming generation.

Implementing blockchain-enabled SDN in resource-constrained devices like internet of things (IoT) presents several challenges. One of the main concerns is the scalability of the current solutions, as they may not be able to manage and secure a large number of nodes without causing severe performance degradation or failures. In addition, the resource-intensive nature of blockchain algorithms may impede the acceptance of this technology in IoT devices, which are often limited by their computing capabilities. Moreover, the performance of the system can be influenced by the blockchain platform chosen, and some of the most suitable platforms may require high costs in terms of implementation, processing, and energy. Balancing power consumption, performance, and security is a significant obstacle in this field.

Implementing blockchain technology in software-defined networks (SDNs) for devices with limited resources requires a challenging approach involving hierarchical architecture and a task-offloading strategy. A hierarchical architecture ensures that the system is well organised and enables effective management of resources. On the other hand, task offloading approaches play a critical role in ensuring the efficient execution of tasks by delegating some of the computing tasks to remote servers. With the increasing demand for secure and efficient deployment of blockchain technology, the need for a hierarchical architecture and task-offloading approaches becomes even more critical. Therefore, implementing these approaches is essential in ensuring the successful deployment of the SDN blockchain in resource-constrained devices.

2. Related Work

The combination of SDN, blockchain (BC), and (IoT) is a field of study that shows potential in delivering effective and protected communication for diverse applications, such as transportation, healthcare, smart cities, and industry. In this review, we examine recent works that leverage the advantages of these technologies to address the challenges of security, privacy, and scalability in IoT networks.

One of the primary advantages of SDN is its ability to separate the data plane and control plane, providing flexible and programmable network management.

In Xie et al. [8] SDN technology was employed to improve the security of 5G Vehicular Ad-hoc Networks (VANETs) by detecting malicious vehicular nodes and messages using a blockchain-based IoT network. Similarly, in Zhang et al. [5], a framework named Block-SDV was suggested for smart cities which is based on a distributed blockchain-enabled SDN VANET. It also incorporates a new approach based on deep Q-learning to enhance the trust features of BC nodes, consensus nodes, vehicles, and the computational capacity of BC. The authors demonstrated the efficacy of their proposed framework and deep Q-learning in a simulation environment.

Gao et al. [6] presented a novel approach that combines Blockchain and SDN for improving security and detecting malicious activities in 5G and fog VANET networks. This integration of BC and SDN has been utilised in the development of Intrusion Detection Systems (IDSs) across various industries, as reported in Derhab et al. [1].

Houda et al. [13] proposed Cochain-SC architecture and suggested two approaches for preventing DDoS (distributed denial-of-service) attacks in SDN, known as intra-domain and inter-domain mitigation. Intra-domain mitigation techniques use entropy-based and Bayes-based methods to classify and reduce the effects of malicious traffic, while inter-domain collaboration is enabled through an Ethereum-based architecture that connects SDN-based autonomous systems to combat DDoS attacks. The evaluation of the proposal considers its effectiveness, flexibility, security, and cost-efficiency.

Sharma et al. [9] proposed an architecture named DistBlockNet that promotes collaboration between SDN and BC in IoT networks. The approach formalises the flow rule table using blockchain, leading to improved performance compared to previous methods. In a similar vein, Rahman et al. [4] introduced a distributed BC-SDN architecture named DistBlockBuilding designed for smart cities [3]. DistBlockBuilding incorporates a cluster-head selection algorithm that efficiently collects sensor data. The performance of DistBlockBuilding was evaluated using various parameters, including latency, throughput, and packet arrival rate.

Chaudhary et al. [11] aimed to utilise the advantages of BC technologies in transportation networks while incorporating SDN to provide authorised resources with low power consumption. At the same time, Li et al. [14] proposed a secure and decentralized framework that integrates both SDN and BC to ensure efficient, reliable, and sustainable energy management. The framework is designed to enhance the security and reliability of energy management while optimiszing energy usage.

In recent studies, fog environments have been a specific area of interest for researchers aiming to develop secure IoT-based applications. Muthanna et al. [2] propose a fog-based IoT system that combines SDN and BC technologies to provide increased privacy, security, and availability. The integration of SDN allows for better management of network traffic, while BC ensures decentralisation in a secure manner Muthanna et al. [2]. To evaluate their architecture, they analysed various factors such as latency, network efficiency, and resource utilisation. In a different study, Sharma et al. [10] presents a unique architecture based on fog computing, SDN, and BC technology. In their approach, fog nodes act as SDN controllers situated at the network's edge, and they present an architecture aimed at guaranteeing high availability. Lastly, hierarchical architecture is a widely adopted approach for designing complex systems because of its numerous advantages.

In terms of security, a hierarchical architecture provides a more robust defence mechanism against attacks as it segregates different levels of access and privileges. This helps in preventing unauthorised access to critical resources and sensitive information. Additionally, the modular structure of the hierarchical architecture allows for easier management of security policies and procedures, which can be customised at different levels.

Regarding efficiency, the hierarchical architecture enables task offloading, which is the process of distributing tasks among different layers of the architecture. This approach helps in reducing the load on the individual layers, leading to improved performance and faster response times. It also enables resource sharing and optimisation by allowing resources to be allocated efficiently across the different layers.

In conclusion, hierarchical architecture is a preferred design approach for complex systems, particularly those requiring a high level of security and efficiency. By segregating access and privileges at different levels and allowing for task offloading, it helps in providing a more robust defence mechanism against attacks and improving system performance. In the next section, we will discuss some of the key considerations when implementing a hierarchical architecture for secure and efficient deployment.

3. System Architecture

3.1 Hierarchical Architecture for Secure and Efficient Deployment [3]

3.1.1 Hierarchical Architecture for SDN Blockchain in Resource-Constrained Devices

The SDN-IoT ecosystem with blockchain integration is organised in a hierarchical architecture that encompasses three layers, each pertaining to a different environment (Fig. 1). The first layer, known as the Perception Layer, is responsible for real-time data sensing and transmission and is composed of IoT sensors and devices. The next sublayer of the Perception Layer selects cluster heads (CHs) to receive data from forwarding devices, such as phones, switches, and routers, and forwards all sensor information to the SDN Environment, which is the second layer of the architecture. Within the Edge Layer of the SDN Environment, there are two levels, the data plane and control plane, which allow IoT devices to transmit data via dynamically managed SDN-IoT gateways using the OpenFlow protocol by multiple SDN controllers. The topmost layer in the proposed architecture is the Cloud Layer, which includes the Blockchain Environment and data centres. This layer serves as a means of data transmission and storage, and its functionality is enhanced by cloud computing, which provides a real-time shared database. The use of blockchain technology in this layer provides added security, privacy,

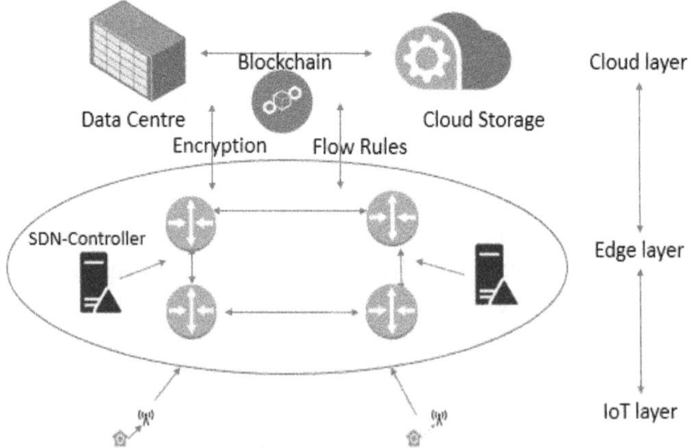

Fig. 1. Proposed system architecture.

and confidentiality for data blocks by ensuring effective communication between them through its chain structure.

In summary, the main objective of the hierarchical arrangement in the SDN-IoT ecosystem is to optimise resource management in the IoT network by utilising an SDN framework integrated with Blockchain technology. The architecture comprises of three distinct layers, specifically tailored to cater to the requirements of each environment, including the IoT Environment, the SDN Environment, and the Blockchain Environment. The Perception Layer, Edge Layer, and Cloud Layer play a pivotal role in the architecture and work together to secure and efficiently deploy the task-offloading strategy and manage resources in the IoT network.

3.1.2 Benefits of the hierarchical architecture in terms of scalability, security, and efficiency

The hierarchical architecture of the Blockchain-enabled SDN-IoT ecosystem offers several benefits in terms of scalability, security, and efficiency.

3.1.2.1 Scalability

The hierarchical architecture provides a scalable solution for managing the increasing number of IoT devices in the network. With the Perception Layer, Edge Layer, and Cloud Layer, the architecture offers a distributed and scalable approach for managing IoT devices and data. The Perception Layer, which includes IoT sensors and devices, allows for the collection and transmission of data to the Edge Layer. The Edge Layer, which includes the SDN environment, can dynamically manage and filter the IoT devices' data through SDN common gateways. Finally, the Cloud Layer offers a real-time shared database and blockchain environment, providing scalability for the network.

3.1.2.2 Security

The hierarchical architecture provides a secure solution for managing IoT devices and data. With the Perception Layer, Edge Layer, and Cloud Layer, the architecture offers a distributed and secure approach for managing IoT devices and data. The Perception Layer provides the first level of security by selecting the CHs with higher energy, ensuring data is transmitted to a secure location. The Edge Layer, which includes the SDN environment, offers security by managing and filtering the

IoT devices' data through SDN's common gateways. Finally, the Cloud Layer provides extra security, privacy, and confidentiality among the data blocks through the blockchain environment.

3.1.2.3 Efficiency

The hierarchical architecture provides an efficient solution for managing IoT devices and data. With the Perception Layer, Edge Layer, and Cloud Layer, the architecture offers a distributed and efficient approach for managing IoT devices and data. The Perception Layer ensures efficient resource management by selecting CHs with higher energy, optimising data transmission. The Edge Layer offers efficient management and filtering of the IoT devices' data through SDN's common gateways, optimising network traffic. Finally, the Cloud Layer provides real-time shared database and Blockchain environment, offering an efficient solution for managing data in the network.

In summary, the hierarchical architecture of the Blockchain-enabled SDN-IoT ecosystem offers several benefits in terms of scalability, security, and efficiency, providing a distributed and optimised approach for managing IoT devices and data.

3.2 Task Offloading Strategy for Improved Efficiency [7]

3.2.1 Task Offloading Strategy for SDN Blockchain in Resource-Constrained Devices

To start up the system, the SDN controller initiates the creation of a database that contains resource information for each fog node through inquiry-response communication. This database is designed to store various information about the fog nodes, including their remaining bandwidth, computational capacity, transmitting power, location, queued tasks, allocated CPU resources, remaining storage, and background noise power. The SDN controller also sets up a database containing edge device information that includes details such as the transmitting power, location, and trustworthiness of the edge devices that have been evaluated by the fog nodes. The controller periodically requests updates from the fog nodes and healthcare IoT edge devices to keep the system's information up to date.

When an edge device aims to offload a task, it transmits a message comprising of the task deadline, price index, and an emergency indicator variable to a broadcast. The received signal strength is measured by each fog node upon getting the message, and an ECHO message, including supported task scheduling policy and service price, is returned to the edge device. The scheduling strategy is chosen depending on the delay requirements of the task, either first-come-first-served or pre-emptive scheduling. The edge device creates a viable set of fog nodes, eliminates nodes with signal strength below the minimum threshold, and communicates with the SDN controller through a REQ message to obtain fog node assignment for task processing.

The edge device's blacklist status is verified by the controller through the database and then requests information updates from the fog nodes. The target fog node for task offloading is determined using the LSRDM-EH centralised decision-making algorithm. To accomplish this, the controller creates a smart contract named SCT1, which includes the flow table, and broadcasts it among TopChain peers. The flow table is then distributed to the relevant fog nodes once consensus is achieved, and SCT1 is safely stored.

3.2.2 Benefits of Task Offloading in Terms of Energy Efficiency, Reduced End-to-End Delay, and Improved Reliability

Task offloading can offer a range of benefits for mobile devices in terms of energy efficiency, reduced end-to-end delay, and improved reliability. By offloading computationally-intensive tasks to more powerful resources, such as cloud servers or edge devices, mobile devices can conserve energy and prolong battery life. This is because offloading reduces the workload on the mobile device's processor, which consumes a significant amount of energy.

Furthermore, offloading can reduce end-to-end delay by leveraging the computational power of remote resources that may be more capable of processing tasks quickly. Efficient management of network traffic is critical for applications that need to process data in real-time, such as autonomous driving or augmented reality. The SDN with BC technology has emerged as a viable solution, as it allows separation of the control and data planes, enabling more efficient and flexible management of network traffic. By offloading tasks to resources with lower latency, end-to-end delay can be minimised, resulting in a more seamless user experience.

Finally, task offloading can improve the reliability of mobile devices by reducing the risk of system crashes or failures due to overloading the mobile device's processor. By offloading tasks to more powerful resources, the mobile device's processor is not overburdened, reducing the risk of system crashes or failures. A more robust and dependable system can be achieved, which is essential for critical applications such as medical or emergency response.

Overall, task offloading can offer significant benefits for mobile devices, including improved energy efficiency, reduced end-to-end delay, and improved reliability. As such, it is an important strategy to consider for developers seeking to optimise the performance and functionality of mobile applications.

3.2.3 Description of the Centralised Algorithm for Low-Latency, Secure, and Reliable Decision-Making

The process of offloading resource-intensive computations to the fog layer for immediate and responsive data processing is made possible by the algorithm after signalling interactions.

The implementation of the algorithm involves multiple steps that are executed by the SDN controller. At first, the controller computes the channel gain between the fog nodes and the edge device using the distance between them and the small-scale attenuation coefficient. Next, it verifies if the total time taken by each fog node to process the task is less than the task deadline. If the total time cost exceeds the task deadline, the fog node is excluded from the feasible node-set.

Afterwards, the controller looks for the N lowest total time cost values and modifies the workable node group with the N corresponding fog nodes. Subsequently, it retrieves the reputation value of each fog node in the workable node group from the database and standardises each element. The controller calculates the fitness function of every fog node in the workable node group by allotting weight parameters to each decision factor.

Ultimately, after evaluating the fitness function, the controller selects the most suitable fog node as the target and creates a flow table that includes the optimal target fog node, the edge device, and the remaining N–1 feasible fog nodes. This flow table is then disseminated to the relevant nodes, including the edge device, the chosen fog node, and the remaining feasible nodes to be prepared as candidate nodes.

The algorithm that is centralised in nature has multiple advantages. The foremost benefit is that it offloads computationally intensive tasks to the fog layer which ensures low latency and real-time data processing. Secondly, the reliability of the system is enhanced by eliminating the fog nodes from the list of feasible nodes if their total time cost exceeds the task deadline. Lastly, to provide security to the system, the reputation value of each fog node is retrieved from the database, and the fitness function of each fog node is calculated based on the assigned weight parameters. In summary, the centralised algorithm facilitates secure, reliable, and low-latency decision-making for healthcare IoT edge devices.

3.3 Encryption Method for Inserting Data into Blocks [12]

To strengthen the security of SDN controllers in fog and IoT environments, the proposed solution aims to enhance security in the SDN architecture by incorporating a paBC network, assuming that the fog nodes act as SDN controllers and establish flow rules for IoT users and other fog nodes through networking operations and routines. It is also assumed that the SDN controllers are connected to each other through east-west communication.

To enhance the security of SDN controllers in fog and IoT environments, a public blockchain (paBC) network is utilized. Fog nodes in this network establish flow rules and transmit them to predetermined miners. Suppose that at timestamp t_a, a controller c sets a flow rule set F_t^{Cx} for application X. If the controller is compromised (c_X^{atk}) at timestamp $t_a + 1$ and sets flawed flow rules $F_{t_a}^{c_X^{atk}}$ for the paBC, these errors can be detected by the miners. If any anomalies are detected, the system remains unchanged as the miners will not create a block for $F_{t_a}^{c_X^{atk}}$. However, if new blocks are created for $F_{t_a}^{c_X^{atk}}$ and the fog nodes adopt these rules, the cloud servers can identify the error and instruct the SDN controllers to revert to the block with flow rule $F_{t_a{}_X^c}$. The paBC architecture ensures that flow rule storage is decentralised and data integrity is maintained through the property of immutability.

The proposed solution for retracting to previously set flow rules in the blockchain involves a three-step process. In step 1, the cloud servers detect the anomaly and suggest retracting to a previous block. The required block may not be the immediately preceding one since the number of applications running in the paBC affects the retraction. In step 2, the retraction occurs, and the fog nodes adopt the previously set flow rules. Finally, in step 3, the cloud servers command the miners to remove the block $F_{t_a}^{c_X^{atk}}$. In the future, the goal is to enable the fog nodes to autonomously detect network stalling and denial of services.

To improve the security of the system, it is recommended to encrypt the flow rules prior to their insertion into the blocks. This is particularly important since the data obtained from IoT devices may be sensitive, and unauthorised access may result in severe consequences. Encryption of the flow rules using algorithms like AES-128, AES-256, or RSA can help prevent undesired access. However, the selection of the encryption technique may depend on the configuration of the device and the policies related to key exchange.

3.3.1 Benefits of Encryption in Terms of Data Security and Confidentiality

This approach offers several benefits in terms of data security and confidentiality. Firstly, encryption ensures that only authorised parties can access the data, even if the data falls into the wrong hands. In the case of the SDN blockchain, encryption would prevent unauthorised entities from accessing and manipulating the flow rules, which are crucial for ensuring network security and efficiency. Encryption can help prevent data breaches and unauthorized access, which could compromise the network and its users.

Secondly, encryption enhances data confidentiality by ensuring that the data remains secure and private. This is particularly important in the case of sensitive data, such as personal information, financial data, and intellectual property. In the context of the SDN blockchain, encryption would ensure that the flow rules remain confidential and secure, preventing any unauthorised access or tampering. This would help maintain the integrity and reliability of the network, as well as build trust and confidence among its users.

Thirdly, encryption ensures data integrity by preventing unauthorised modifications to the data. Encryption can help detect and prevent any attempts to modify or tamper with the flow rules, thereby ensuring that the network operates as intended. This helps prevent disruptions and attacks on the network, which could result in data loss, downtime, and other security risks.

Overall, the use of encryption in the SDN blockchain can enhance data security and confidentiality, thereby ensuring that the network operates in a secure and reliable manner. By encrypting the flow rules, the network can prevent unauthorised access, maintain data confidentiality, and ensure data integrity, which are crucial for building trust and confidence among its users.

3.4 Flow-Rules in SDN Blockchain to Ensure a Tamper-Proof Record of the Rules Enforced in the Switches [12]

The use of blockchain technology is proposed in SmartBlock-SDN to improve resource management in IoT networks. The proposed solution for enhancing the security of SDN architecture involves the integration of two separate blockchains. The first blockchain is responsible for managing distributed flow rules and is updated by the system administrator using a REST API. This blockchain functions as a version control system that logs all changes to ensure consistency among the controllers' rule set. On the other hand, the second blockchain is responsible for verifying flow rules, detecting rouge nodes, and dumping flow rules sequentially. If a switch fails to dump the same rules, it is isolated from the network. Additionally, a stealthy intrusion detection system (IDS) is integrated into the SmartBlock-SDN framework to detect flow-rule violations and switch isolation.

3.4.1 Benefits of Enforcing Flow Rules in Terms of Network Security and Tamper-Proof Record Keeping

In terms of security, enforcing flow rules ensures that switches in the network are only allowed to send and receive data based on specific criteria that are defined by the controller. This can prevent unauthorised access and malicious attacks from rogue switches or adversaries. By using blockchain to maintain the consistency of the flow rules, any attempt to modify or manipulate the rules will be immediately detected, and the switch in question will be isolated from the network.

Furthermore, the use of BC technology for record-keeping purposes ensures that all updates to the flow rules are recorded in an immutable ledger. This creates a tamper-proof record of all flow rule changes, which can be used to verify the integrity of the network and identify any potential issues or breaches. Additionally, this ledger can be used to trace the history of flow rules and identify when and why specific rules were added or modified, which can be helpful for troubleshooting and future planning.

Overall, enforcing flow-rules through the use of BC technology can enhance network security, provide tamper-proof record-keeping, and ensure the integrity and consistency of the network's flow rules.

4. Conclusion

In conclusion, the proposed hierarchical architecture, task offloading strategy, data encryption, and flow-rule enforcement offer a promising solution for the secure and efficient deployment of SDN blockchain in resource-constrained devices. These methods can significantly improve the performance, scalability, and security of blockchain-based SDN networks, making them more suitable for deployment in IoT and other resource-constrained devices. The potential impact of these proposed methods on the adoption of blockchain-based SDN in resource-constrained network devices is substantial. With the increasing demand for secure and efficient networking solutions in IoT and other resource-constrained devices, the proposed methods can enable the deployment of blockchain-based SDN networks in these devices, facilitating the development of new applications and services. Future research in the field of SDN blockchain in resource constrained devices should focus on exploring new techniques to further enhance the security and performance of SDN blockchain networks. This could include developing new algorithms for task offloading and flow-rule enforcement, improving the efficiency of data encryption, and exploring new consensus mechanisms for blockchain-based SDN networks. By addressing these research challenges, we can further advance the field of SDN blockchain and enable the development of new applications and services for resource-constrained devices.

References

[1] Derhab, A., Guerroumi, M., Gumaei, A., Maglaras, L., Ferrag, M.A., Mukherjee, M. and Khan, F.A. 2019. Blockchain and random subspace learning-based IDs for SDN-enabled industrial IoT security, *Sensors*, 19(14): 3119.

[2] Muthanna, A., Ateya, A.A., Khakimov, A., Gudkova, I., Abuarqoub, A., Samouylov, K. and Koucheryavy, A. 2019. Secure and reliable IoT networks using fog computing with software-defined networking and blockchain. *Journal of Sensor and Actuator Networks*, 8(1): 15.

[3] Rahman, A., Nasir, M.K., Rahman, Z., Mosavi, A., Shahab, S. and Minaei-Bidgoli, B. 2020. Distblockbuilding: A distributed blockchain-based SDN-IoT network for smart building management. *IEEE Access*, 8: 140 008–140 018.

[4] Rahman, A., Islam, M.J., Montieri, A., Nasir, M.K., Reza, M.M., Band, S.S., Pescape, A., Hasan, Sookhak, M. and Mosavi, A. 2021. Smartblocksdn: An optimised blockchain-SDN framework for resource management in IoT. *IEEE Access*, 9: 28 361–28 376.

[5] Zhang, D., Yu, F.R. and Yang, R. 2019. Blockchain-based distributed software-defined vehicular networks: A duelling deep Q-learning approach. *IEEE Transactions on Cognitive Communications and Networking*, 5(4): 1086–1100.
[6] Gao, J., Agyekum, K.O.-B.O., Sifah, E.B., Acheampong, K.N., Xia, Q., Du, X., Guizani, M. and Xia, H. 2019. A blockchain-SDN-enabled internet of vehicles environment for fog computing and 5G networks. *IEEE Internet of Things Journal*, 7(5): 4278–4291.
[7] Ren, J., Li, J., Liu, H. and Qin, T. 2021. Task offloading strategy with emergency handling and blockchain security in SDN-empowered and fog-assisted healthcare IoT. *Tsinghua Science and Technology*, 27(4): 760–776.
[8] Xie, L., Ding, Y., Yang, H. and Wang, X. 2019. Blockchain-based secure and trustworthy internet of things in SDN-enabled 5G-vanets. *IEEE Access*, 7: 56 656–56 666.
[9] Sharma, P.K., Singh, S., Jeong, Y.-S. and Park, J.H. 2017. Distblocknet: A distributed blockchain-based secure SDN architecture for IoT networks. *IEEE Communications Magazine*, 55(9): 78–85.
[10] Sharma, P.K., Chen, M.-Y. and Park, J.H. 2017. A software defined fog node based distributed blockchain cloud architecture for IoT. *IEEE Access*, 6: 115–124.
[11] Chaudhary, R., Jindal, A., Aujla, G.S., Aggarwal, S., Kumar, N. and Choo, K.-K.R. 2019. Best: Blockchain-based secure energy trading in SDN-enabled intelligent transportation system. *Computers & Security*, 85: 288–299.
[12] Misra, S., Deb, P.K., Pathak, N. and Mukherjee, A. 2020. Blockchain-enabled SDN for securing fog-based resource-constrained IoT. pp. 490–495. *In*: *IEEE INFOCOM 2020-IEEE Conference on Computer Communications Workshops (INFOCOM WKSHPS)*. IEEE.
[13] Abou El Houda, Z., Hafid, A.S. and Khoukhi, L. 2019. Cochain-sc: An intraand inter-domain DDOS mitigation scheme based on blockchain using SDN and smart contract. *IEEE Access*, 7: 98 893–98 907.
[14] Li, Z., Shahidehpour, M. and Liu, X. 2018. Cyber-secure decentralized energy management for IoT-enabled active distribution networks. *Journal of Modern Power Systems and Clean Energy*, 6(5): 900–917.

Chapter 3

Computer Vision Mapping for Parking Space Counter

Atul Srivastava and *Abdul Azeem Khan**

1. Introduction

The era people are living is very automated and dependent upon the use of technology for every small thing, be it a toothbrush or keeping ourselves warm. As humans have a tendency to be in their comfort zone and make things easy for themselves, or else feel lazy to work or get frustrated very easily. For transportation, most of the people find it really convenient to have their own vehicles rather than using a public transport which is quite slow and takes considerable time. This has resulted in a many people buying their individual vehicles which have thus increased in large numbers. As a result, endless problems are occurring such as innumerable individuals are parking their vehicles on the roadside as there is no proper parking management system. Traffic increases and takes a lot of time resulting in people getting agitated and frustrated when driving on jam-packed roads and not reaching their destination on schedule. When it comes to parking, there is no proper place where

Amity School of Engineering and Technology, Amity University Uttar Pradesh (Lucknow).
Email: asrivastava5@lko.amity.edu
* Corresponding author: abdulazeemkhanofficial@gmail.com

drivers can find empty spaces and park their vehicles. Distributed parking systems take a plenty of time for the drivers to find the spots for parking after managing and shuffling a lot of cars in the parking lot manually and deciding who arrived first and who needs to leave the lot next.

Most of the parking systems that are already present utilise sensors for the parking but its time-consuming because one sensor would work only for a particular vehicle or a single parking space. The cost of managing and maintaining such systems is quite typical and requires a lot of monitoring. This paper proposes to develop a parking system which can easily work using cameras [18], and processing the images generated and applying them to the real-life footage using open Computer Vision [11].

Computer Vision is a variety of artificial intelligence which is first applied on the images and further on videos or camera footages after being activated on the images. It is a technique which is used to detect the empty parking spaces using the vision provided by the Open CV (computer vision) library [3]. Computer vision works on images and videos and takes data input and processes to take some actions. Computer vision is similar to human vision as it can detect and perform necessary functions but has some limitations when compared with human vision. Unlike human vision it needs to read data from a camera or other input sources and then learn from the data and perform operations in real-time situations. The approach proposed above can be implemented by the addressing the points mentioned below:

Step1: Trying to develop a web-based application that the users for the parking system need to register and sign in to the system to use the application conveniently. They must provide the basic information about themselves and their requirements. Further customised modifications can be done to meet the needs of the parking users. Figure 1 shows the login page.

Step 2: The developer would be working according to the requirements of the users. It is like a school kind of model where identifying which student belongs to which class is easy. So that the application accordingly fulfils the wants of the registered users. Figure 2 represents the template for requesting a parking space.

Figure 2 shows different boxes to fill where a user can specify name, location, size of parking lots, footage, and addresses.

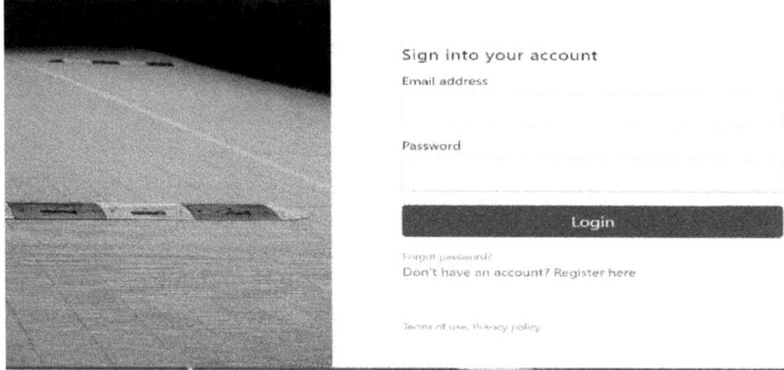

Fig. 1. Login page for the users to sign in or login.

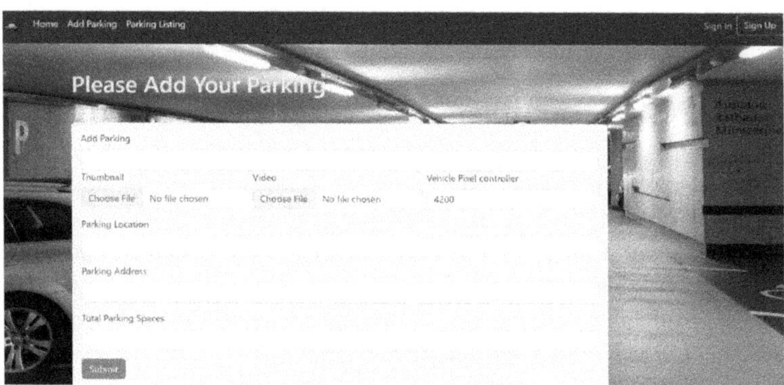

Fig. 2. Parking request template.

Step 3: This includes the collection of the clips and camera footage and to send it to the systems so that it can be easily processed and results could be given on the input data as shown in Fig. 3.

In the Fig. 3, observation is done for all the parking information added by different users, users can view each of them by opening them and after using they can delete them as well. Figure 4 shows the information about a particular parking and users can use the camera view and use the editor to edit the parking spaces.

Step 4: After processing of the input data, results would be given about which parking slots are empty out of the total parking spaces. All the empty and non-empty can be easily identified by the systems using a colour coding scheme say green for an empty slot and red for a slot

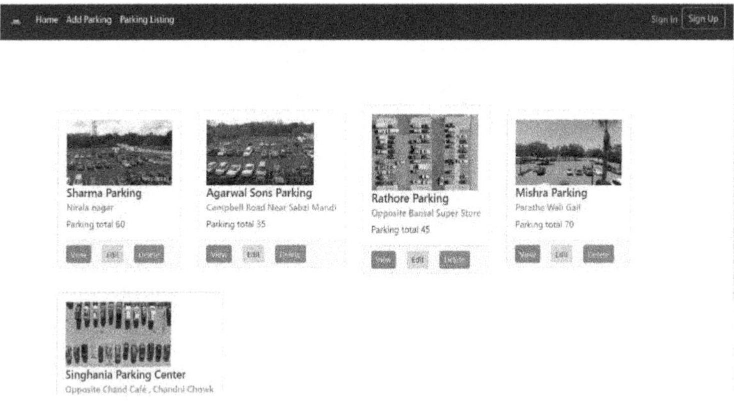

Fig. 3. Parking listing page when a user adds information.

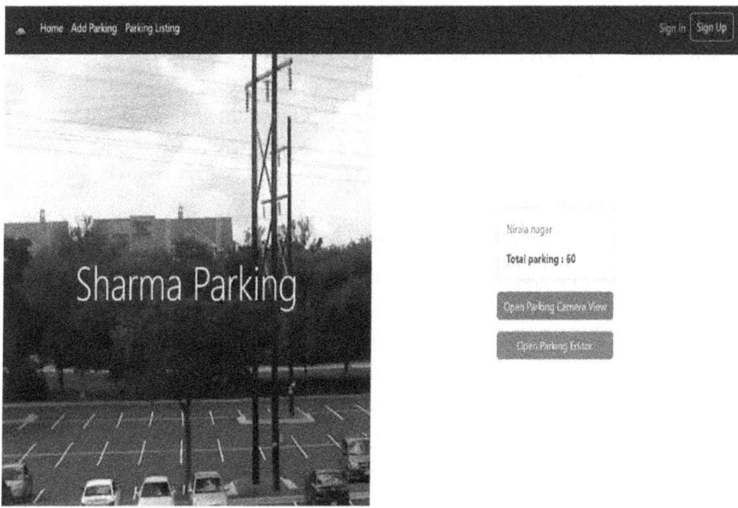

Fig. 4. Information about a particular parking lot.

which is already occupied by some other user. The count of the free spaces would also be displayed in the systems for smooth and better management as shown in Fig. 5.

These steps can be used very easily to get started with the parking system.

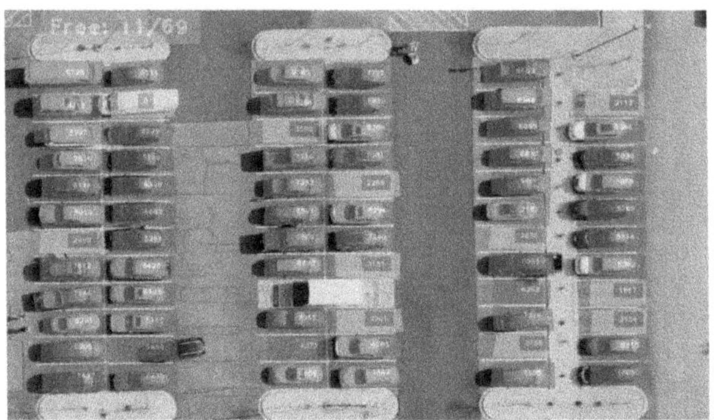

Fig. 5. Free parking spaces after processing (marked as light Grey).

2. Related Work

The advancement of computer vision for smart parking and space identification has drawn more attention in recent years. This has been sparked by the rise of urban parking space demand as well as the various benefits of more effective parking management, such as decreased traffic and pollution as well as higher profits for parking lot owners. The implementation of computer vision in smart parking, including parking spot recognition and occupancy prediction, has been examined in a number of research. For instance, Zhou et al. [25] created a deep learning-based car park recognition system that found parking spaces in an inside parking lot with high accuracy.

In our research and finding there is an observation that there are generally two methods that are used for car parking detection and they are quite different one another. Trying to learn about both of them and compare and find which one of them is the best and which are the new methods that are proposed in recently published research papers. The first method is a process in which checking whether a particular space it is occupied or not and can be said as a space-type parking system. While on the other hand the other methods check if there is a car parked in a given region and can be said as a car- or vehicle-type parking system. There have been proposals about modifying these systems so that not only the occupancy of the vehicles can be found but also the size and position can be included in the features which will eventually help us to eliminate the vehicles which are not properly parked in the parking spaces defined in different shapes in different parking spaces.

Some of the researchers have also proposed parking management systems which include artificial intelligence, sensors, and data communication and this digital data while communicating can be in the form of video, CCTV footage, or camera footage. This type of process not only applies to parking only, but there are other methods which can be applied to the traffic management but that's an out-of-scope context for us. These types of traffic systems generally use sensors that are good quality and can detect the cars in the parking lots. The next step is to take the input by using image processing and these images are further improved in quality so that the number plates of the vehicles can be easily recognized and each character can be processed into the database. The information is present in the database, and data for incoming cars can be added and for outgoing cars can be deleted. According to the cars parked for a given period of time which is already present in the database, the cost of parking can be calculated and customers can be billed accordingly. The best thing about this type of process is that it is quite fast and there is less latency between incoming and outgoing traffic. The system was able to find parking spaces in real-time with a 95% accuracy by combining convolutional neural networks (CNNs) plus image processing methods.

Similar to this, Wang et al. [15] described a parking detection method that employed a mixture of deep learning and machine vision methods to precisely detect and categorise car parks in outdoor settings. The system identified parking spaces with 91.5% accuracy in a sizable outdoor parking lot using a deep convolutional neural network (DCNN) to recognise their boundaries.

Other research has concentrated on the application of computer vision to the assessment of occupancy, or the number of cars filling a parking space. To predict the number of vehicles in each parking place in real-time, Zhang et al. [24] proposed a deep learning-based occupied estimation system that combined CNNs with recurrent neural networks. With an absolute mean error of only 0.17 cars per space, the system achieved great accuracy.

A rising number of people are also interested in using mapping technologies, like LiDAR and three-dimension imaging, for parking spot detection and mapping, in addition to these investigations. For instance, Yoo et al. [19] suggested a LiDAR-based carpark recognition system that was capable of recognising parking spaces at a variety of angles and distances and obtained excellent accuracy in identifying parking spots in outdoor contexts. The ability with computer vision and mapping technology for parking spot detection and management is overall shown by these works. By creating a cutting-edge computer vision and method that can precisely identify and count parking spots in real-time and interact with current parking control systems to enhance efficiency and revenue production, the research proposal aims to improve upon this previous work.

A vehicle park detection system was created by Islam et al. [8] using Python and OpenCV to find cars in a parking lot. The technology was capable of identifying autos in real-time and had a detection accuracy of 98.5%. The research proposal will seek to expand on this work by creating a unique model that can precisely detect and count vacant parking spaces that use a combo of computer vision and process. This research demonstrates the capability of OpenCV and Python for car park management and detection. Deep reinforcement learning was applied by Kim et al. [9] to create an indoor parking space recognition system that had a 97.5% accuracy rate. With an efficacy of 92.5%, Chen et al. [5] detected outdoor parking spaces using YOLOv3 and SSD. With an efficiency of 94.6%, Shao et al. [14] enhanced on YOLOv3 for locating outdoor parking spaces. With an accuracy of 88.3%, Wang et al. [14] used SVM with HOG for the detection of outdoor parking spaces. The

Table 1. This is a comparison of different car parking systems and MAE stands for mean absolute error.

Study	Parking Environment	Detection Method	Accuracy
Zhou et al. [25]	Indoor	CNNs and Image Processing	95%
Wang et al. [15]	Outdoor	DCNNs	91.5%
Zhang et al. [24]	Indoor	CNNs and RNNs	MAE: 0.17 vehicles per space
Yoo et al. [19]	Outdoor	LiDAR	High Accuracy
Islam et al. [8]	Outdoor	OpenCV and Python	98.5%
Kim et al. [9]	Indoor	Deep Reinforcement Learning	97.5%
Chen et al. [5]	Outdoor	YOLOv3 and SSD	92.5%
Shao et al. [14]	Outdoor	Improved YOLOv3	94.6%
Wang et al. [16]	Outdoor	SVM and HOG	88.3%

progress of the suggested research on computer vision mapping for parking space counts can benefit from the additional studies for a wider range of detecting techniques and parking situations.

2.1 Relation and Applications in Blockchain Technology

Yes, there can be a connection between blockchain technology and computer vision mapping in parking spot counting. Blockchain is a decentralised, distributed ledger system that makes it possible to store transaction data securely and openly. There are a number of potential domains where computer vision mapping plus blockchain can interact, even though their direct relationship may not be immediately obvious:

a) *Data Verification and Integrity*: The use of computer vision mapping to count parking spaces depends on accurate and trustworthy parking space data. The acquired data can be recorded in a decentralised ledger using blockchain technology, guaranteeing its authenticity and immutability. This can aid in preventing data manipulation or unauthorised changes, fostering trust and openness in the parking spot counting procedure.

b) *Data exchange and Collaboration*: Blockchain can enable safe and effective data exchange in instances where numerous entities or parties are involved in managing parking spaces. Using blockchain-based smart contracts, other parties like parking lot managers, local government agencies, or autonomous automobiles, can access and change the parking space information. This makes collaboration simple and eliminates the need for middlemen.

c) *Tokenized Incentive Systems*: The development of designated reward systems for parking spot utilisation is possible because of blockchain technology. Parking spot data can be utilised to calculate parking availability, rates of occupancy, and demand trends by merging computer vision modelling with blockchain. This information can subsequently be used to develop flexible price structures or incentive schemes, promoting effective parking space utilisation and easing traffic.

d) *Decentralised Parking Management*: Conventional parking management platforms are frequently centralised, which creates possible data vulnerabilities and single points of failure. Parking spot data can be decentralised and spread across several nodes using blockchain, improving resilience and boosting security. Peer-to-peer transactions can be made possible via this decentralised strategy, enabling people to rent out or trade parking spaces without the use of middlemen.

It's crucial to remember that, despite the possibility of benefits, the use of blockchain and computer vision mapping in parking space accounting may potentially add new complexity and factors. In the planning and execution of such systems, concerns including privacy, capacity, and instantaneous processing of data must be carefully considered.

2.2 Applications in Blockchain Technology

Blockchain technology can be used for parking spot counting using machine vision mapping. Several possible applications are listed below:

a) *Parking space rentals using smart contracts*: Smart contracts operate as self-executing contracts in which the conditions of the agreement are put directly into the programme code. Smart agreements can automatically enable the rental or allocation of parking spaces by employing computer vision modelling to ascertain the availability of parking places. These agreements can specify the terms, duration, and payment schedules, enabling automated and secure transactions between parking lot owners and tenants.

b) *Transparent Parking Spot Management*: Blockchain can be used to give a transparent and verifiable record of parking spot management because of its transparency and immutability. Stakeholders can access a common, tamper-proof source of information by storing parking space data on a blockchain, such as occupancy levels, breaches, and payments. The responsibility and trust between parking businesses, municipalities, and customers can all be enhanced by this transparency.

c) *Informational Data Market for Parking Spaces*: The information gathered through computer vision modelling may be used for purposes other than managing parking spaces. Parking spot data can be shared and monetized securely in blockchain-based data marketplaces. This encourages the collecting of accurate and current parking spot information. Data users, including urban planners, mobility corporations, or researchers, can obtain this data by buying it directly from information providers.

These are just a few instances of how blockchain technology can be used for computer vision imaging for parking spot counting. In the management of parking spaces and related services, the incorporation of these technologies may bring about advantages like increased security, decentralisation, and transparency.

3. Methods and Materials

The goal of the branch of research known as computer vision is to make it possible for machines to comprehend and interpret visual data. It entails the creation of algorithms capable of video or picture analysis and information extraction. A well-liked open-source package for Python computer vision programming is called OpenCV. It provides a variety of features for computer vision and image processing operations. Cao et al. [4] suggested to explore and develop a system that will determine the number of vacant spaces in a car park. To accomplish this it requires utilising a cam that is fixed in the parking area and linking it to a computer. To identify which parking spaces are vacant and which ones are occupied, the computer will examine the video stream coming from the camera and employ computer vision techniques. To begin with, there must be aligning of the camera to remove any visual distortion. The camera calibration features of OpenCV can be used to accomplish this. A checkerboard pattern is photographed repeatedly from various perspectives during the calibrating process to determine the intrinsic and extrinsic properties of the camera. Li et al. [10] opined that it can start identifying the parking places in the image once the camera has been calibrated. To accomplish this, they can threshold the image and retrieve the white lines which delineate the parking space limits. The contours of such white lines can then be discovered using OpenCV's contour detection tools. The outlines must then be divided into separate parking places. By examining each contour's size and aspect ratio, they may accomplish this. Another option is to arrange surrounding contours into a cluster that stand in for specific parking places using a clustering technique.

Once the parking spaces have been divided into sections, they must identify whether or not each place is occupied. By examining the texture and colour of the pixels in each parking place, they may do this. Nguyen et al. [12] proposed a classifier that can differentiate between unoccupied and occupied parking spaces can be trained using machine learning algorithms. Finally, they may determine the quantity of vacant parking spaces by determining the number of places that are designated as such.

A database and a number of datasets can be utilised to manage and store the parking lot photographs and associated data with in computer vision mapping for car park counter utilizing OpenCV and Python. This can be helpful for a variety of things, including saving the photographs for later study, tracking the number of vacant spaces and enhancing the parking space counter's accuracy over time.

Any relational database management systems like MySQL, PostgreSQL, or SQLite can be used to implement a database. The parking lot photographs can be stored in one database, and the associated metadata, such as the time and date of the capturing images, the number of vacant spots, and the number of occupied spaces, can be stored in another table. Python can be used to interface with the database and carry out a number of operations, including uploading new photographs, updating the metadata, including query the databases to get the information they need. The creation and assessment of the machine learning algorithms employed in the parking spot counter depend on the dataset gathering. Wu et al. [17] proposed that they can make their own datasets or use publicly accessible datasets by taking pictures of numerous parking lots in various lighting situations and at various times of the day. To enhance the dataset size and avoid overfitting, they can utilise a variety of strategies, such as data augmentation. Techniques for data augmentation can involve scaling, flipping, rotating, and applying noise to the photos. They can utilise a variety of methods, including supervised methods, unsupervised, and deep learning, to create machine learning algorithms. To create and train the models, they can utilise Python libraries like scikit-learn and TensorFlow. This parking space counter can be created using Python and OpenCV for implementation. All the image processing operations may be carried out using the OpenCV library, and Python can be used to create the code that ties everything together. For further tasks like data processing and visualisation, they can also use additional Python libraries like NumPy, Pandas, and Matplotlib. The calibration of the camera, detection of parking spaces, segmentation of parking spaces, classification of parking spaces, or counting of vacant spaces are all steps in vision-based mapping to parking space counter utilizing OpenCV and Python. The parking lot photos and associated data can be stored and managed with the aid of a databases and a set of datasets. Data augmentation, supervised, unsupervised, and deep learning are some of the processing methods utilised to create the machine learning algorithms. Abadi et al. [2] opined that these methods can be implemented using Python and its different libraries, like scikit-learn and TensorFlow.

4. Experiment and Results

Here's a possible implementation using OpenCV, Python, and SQLite for a parking space counter:

4.1 Data Collection

The first step in creating a parking counter utilising computer vision is data collection. To accomplish this, they will need to take footage or photographs of the parking area they wish to keep an eye on. With a camera, either fixed on a pole over the parking lot or on a moving car, the pictures or video can be taken. The cameras should be placed so they it can clearly see the entire parking lot, including every available space. You must take pictures or record video of the parking lot where you'd like to count the spaces to compile the dataset. To account for varied levels of parking activity, it is advised to take pictures or video at various times of the day and on various days of the week. For the purpose of training the computer vision model, it is also crucial to capture both vacant and occupied parking spaces.

4.2 Image Processing

The following step is to use OpenCV to process the pictures after gathering the photos or videos of the parking area. This entails locating parking spaces in an image or video and figuring out whether or not they are occupied. Use machine learning methods, feature extraction, and object detection approaches to do this. You must pre-process the photos after collecting the dataset to improve their quality and get rid of any extra noise. OpenCV features like picture thresholding, blur, and morphological processes can be used for this. Using a process called image thresholding, a greyscale image is transformed into a binary image, in which each pixel is either white and black depending on a threshold value. The process of blurring involves averaging the image pixels in a specific area to smooth out the image. Erosion and dilation are morphological techniques that are used to fill up the gaps between objects or remove minor amounts of image noise.

4.3 Object Detection and Tracking

Use object detection methods like contour detection or the Hough line transform to the images using OpenCV to find the parking places. A method for locating object boundaries in a picture is contour detection. A method to find straight lines in a picture is the Hough line transform. The parking spaces must then be tracked over a series of frames utilising object tracking algorithms like Kalman filters or optical flow. Computer vision imaging in automobile parking systems, which is revolutionising the way parking spots are controlled and used, is fundamentally based on object identification and tracking. These technologies provide effective parking spot distribution, real-time vacancy monitoring, and enhanced parking management by precisely recognising and tracking automobiles within parking lots. Vehicles entering the camera's area of view are recognised and located by object detection algorithms. These algorithms examine the visual information obtained by cameras positioned in parking lots and identify particular patterns, forms, or colour features to determine the presence of vehicles. Object identification techniques like YOLO (You Only Look Once) with SSD (Single Shot MultiBox Detector) are reaching exceptional accuracy and real-time performance thanks to developments in deep learning as well as convolutional neural networks. The data produced by the identification and tracking of objects algorithms can also be used by parking managers and administrators to study parking habits, optimise space allocation, and put preventive maintenance plans into practise. In conclusion, computer vision mapping in parking systems has been transformed by object identification and tracking technologies. These technologies make it possible to identify, track, and map automobiles in parking lots in real-time, which facilitates effective parking planning and improves user experience. Car parking systems may enhance parking spot allocation, deliver precise occupancy data, and make a difference to smarter and environmentally friendly urban settings by utilising the potential of deep learning as well as sophisticated tracking algorithms.

4.4 Parking Space Counter

Its last step is to tally how many parking spaces are available and occupied inside the parking lot. Analysing the information that was obtained in the previous stage will enable this. There are many technologies that can be used to implement the parking space counter, including Python, SQLite, and other programming languages and databases. The counter may be made to offer real-time updates that can be seen on a web dashboard or mobile application. You may count the number of vacant and full parking spaces after finding and monitoring the parking spaces. This can be achieved by comparing the locations of each parking space that has been detected with those of the parked vehicles. Setting a threshold space between the automobile and the parking place is an easy way to accomplish this. The area is regarded as occupied if the gap is smaller than the threshold. If not, it is regarded as empty.

4.5 Creating SQL Database

You must select a SQL database management system (DBMS), such as SQLite, MySQL, or PostgreSQL, before you can create a SQL database. The file-based, lightweight SQLite database is simple to set up and then use. A GUI tool like DB Browser with SQLite or the command line can both be used to create databases. You must create a table to house all parking space count information after the database has been built. The table should include columns for the day and hour of the count, the quantity of vacant spaces, and the quantity of spaces occupied.

4.6 Inserting Data into the Database

You must utilise a Python library that offers an interface to database management system to be able to enter data into the SQL database. Use Python's built-in sqlite3 module to access SQLite data. You must write a Python script that extracts the computer vision model's parking spot count data and inserts it into the correct SQL database columns.

4.7 Visualizing the Data

Use Python packages like Matplotlib or Seaborn to visualise the data. These libraries offer a variety of tools for building various graphs and charts. You could, for instance, make a line chart that plots the number of parking spaces over time or a bar graphs that contrasts the number of spaces over various days during a week. Moreover, interactive visualisations can be produced using libraries like Plotly or Bokeh.

4.8 OpenCV

OpenCV is a well-known machine vision library that offers a variety of methods for handling still images and moving pictures. It is written in the C++ programming language and includes Python interfaces. You may complete tasks like image filtering, edge detection, extraction of features, object detection and tracking, among others, with OpenCV. You can process the photos or video inputs from the public parking cameras in your research paper regarding computer vision modelling for parking space counter using OpenCV to identify vacant and occupied parking spots. The car park map with the designated parking places can also be visualised using OpenCV. Following the discovery of vehicles, object tracking techniques are used to track their trajectories and movements throughout time. For calculating the spot, velocity, and orientation of vehicles, tracking systems use approaches based on deep learning, optical flow, or Kalman filtering. The system can give real-time information regarding parking spot availability and occupancy by continually modifying the position of the car and tracking its movements. Dynamic parking maps that reflect the current condition of parking spaces can be produced using the detection and tracking of objects together. Users can easily locate available spots for parking and make educated judgements by visualising these maps through online interfaces or mobile applications.

4.9 Python

Python is a well-known high-level language of programming that is simple to learn and includes a variety of frameworks and libraries for different applications, such as computer science, machine learning, and machine vision. Because it includes an interface of OpenCV plus offers simple access to other modules that you might need for your project, Python is an ideal choice for any research study. The duties of parking spot detection and counting can be accomplished using Python by writing the code to analyse the pictures and video feeds from parking lot cameras.

4.10 SQLite

For storing and managing data, SQLite is a compact, fully accessible relational database management system. Because SQLite is simple to use and doesn't necessitate a distinct server process, you can seamlessly connect it into any Python application without having to perform complicated setup, making it an excellent choice on your research paper. Data from the parking lot map, including the number of spaces, their positions, and their occupancy status, can be stored using SQLite. This will give you statistics on how many parking spaces are being used as well as help you monitor the parking lot's condition over time.

Create your user interface (UI) of your web app before you begin the development process. The user interface (UI) should be simple and easy to use. To create the UI, you may employ a prototyping programme like Adobe XD or Figma. The creation process can begin as soon as the UI layout has been finalised. Establish a development environment on your local workstation, set up an interactive development environment. To create the web application, Python as well as a web framework like Django or Flask are required. Install the required software and libraries, including SQL alchemy, NumPy, and OpenCV. Put this computer vision algorithm into practise: Work on the computer vision algorithm with Python and OpenCV. The algorithm ought to have the ability to estimate the number of parking spaces that are available in the image or video that has been provided. Database connection Connect your Python code to a SQL database you've created. To communicate with the database, use SQL alchemy. Create the tables you'll need to record information on parking spots, such as the ID of the parking lot or space, its availability, etc.

4.11 Creating the Website Application

Design the web application using a web framework like Django or Flask. The following attributes ought to be present in the web application:

a) *Choose a Web Framework*: Django, Flask, and Pyramid are a few of the well-liked web frameworks for Python. You can select the framework that best suits your needs from a variety of features and tools provided by these frameworks for developing web apps. It is crucial to choose the best web structure for computer vision projection in parking systems since it has a significant impact on the efficacy and efficiency of the total solution. The core for managing and visualising the massive volumes of data produced by computer vision technologies in real-time parking scenarios is a well-designed and strong web framework. The web framework's scalability and performance must be taken into account first. A trustworthy framework should be able to manage an immense amount of information handling and analysis given the growing number of cars on the road. This includes the mapping of parking places and the detection and tracking of objects in real-time. To ensure smooth operation, it's critical to select a framework that can effectively handle these computationally demanding jobs. The web framework's adaptability and extensibility should be taken into account. Since computer vision technology and algorithms are always changing, the framework of choice should make it simple to include new methods. This makes sure that as automated vision and imaging technologies evolve, the parking system will modify and get better over time. Additionally, a flexible framework enables designers to modify and adapt the solution to particular needs, improving its efficiency and usability. The choice of web framework should come with a wide range of tools and libraries, which is another crucial factor. These technologies can speed up the creation process, make computer vision algorithms easier to deploy, and offer pre-made elements for visualisation and user interaction.

b) *Create a Database*: You'll need to create a database to hold data about parking spaces plus their occupancy. You can utilise a variety of relational database management platforms (RDBMS), including MySQL, PostgreSQL, and SQLite. To specify the layout of your database, you must also design a database schema. Creating an effective and reliable database is a key component in creating an automated vision mapping application for parking. Instantaneous mapping of spaces for parking and effective information retrieval are made possible by the database, which acts as the core for storing and handling the enormous volumes of data produced by computer vision algorithms. Here, they'll go through the main factors to think about and the procedures for building a database for automated vision mapping in parking. In conclusion, considerable thought must be given to database design, data collecting, insertion, storage, the use of indexing security, backups, recovery, and performance optimisation while building the database for computer vision modelling in parking. Developers may create a solid and effective database that serves as the foundation of an advanced auto parking system by following these guidelines and best practices. This will enable real-time tracking of parking spaces, improve user experience, and contribute to smarter urban environments.

c) *Incorporate OpenCV with Your Web-based Application*: You must integrate OpenCV into your preferred web framework to use it in your online application. Downloading the OpenCV library is usually required, then a Python script is made to use OpenCV to find parking spots and count occupancy. To store the outcomes of your parking spot counter, you must additionally establish a connection to your database. The possibilities of computer vision mapping in parking systems can be greatly improved by combining OpenCV (Open-Source Computer Vision Library) into your web-based application. OpenCV is the best option for adding computer vision capabilities since it offers a large variety of potent algorithms and programmes for image manipulation, object detection, and tracking. You can use OpenCV's broad feature set to carry out operations like vehicle detection, licence plate identification, and parking space mapping by integrating it into your web-based application. These features provide the system the ability to precisely determine available parking spaces, automatically recognise and track vehicles inside parking zones, and give users real-time visual feedback.

d) *Create a User Interface*: After setting up your backend, you must do the same for your web-based application. This usually entails using HTML, CSS, and Js to create web pages, as well as a web framework's template engine to show data from your databases and OpenCV code. For a system to be effective and user-friendly, the user interface for machine vision mapping in parking spaces is essential. Users engage with the system primarily through the interface, which gives them the ability to successfully comprehend and make use of the complicated data produced by computer vision algorithms. The user engagement and overall effectiveness of the parking network are both improved by an interface that is well-designed. Spaces for parking and their available status should be represented in the user interface in a way that is both obvious and visually appealing. Information about vacant and occupied areas can be rapidly communicated by using colour-coded indications or icons. Users can better understand the parking lot's layout and spot open spaces thanks to clear flags and visual clues. To provide consumers with correct information, the interface must additionally refresh in real-time to show changes in occupancy. The user interface can be made more interesting by using interactive elements like zooming, panning, and rotating. These features enable visitors to look about the parking lot, inspect particular areas or parking places, and get a full picture of the occupancy levels. Users can seek parking spaces based on certain criteria by using filters or search options, which will speed up the decision-making process. When viewers hover over or select on parking spaces, tooltips or pop-ups may appear with more information. This data may include characteristics like size, features for people with disabilities, or specific designations. On the basis of these facts, users can choose the best parking space for their need. Developing an attractive and instructive depiction of spots for parking and their empty status is a key component of building a user experience for computer vision imaging in car parking. Zooming and filtering are two interactive elements that improve user experience while tooltips offer more information. The user experience can be further improved by additional types of opinions, such as auditory notifications, and the interface's responsiveness and device adaptability. User demands and preferences are met by the interface thanks to testing for usability and iterative design. Developers may design an easy-to-use interface that optimises parking experiences, promotes space management, and raises user happiness by taking these considerations into account.

e) *Deploy your Web Application*: Your web application must be deployed to a web application or cloud platform once you have tested it in a local development setup for people to access it. Python online applications can be deployed on a variety of platforms, including Amazon Web Services (AWS), Microsoft Azure, and Google Cloud. It is essential to implement a web-based programme for machine vision mapping in automobile parking systems in order to put this cutting-edge technology to use in real-world applications. It entails making the solution user-friendly, ensuring that it runs without a hitch, and maximising its capacity to improve parking space management. Discussion of the main factors and procedures for installing such an application in this conclusion. First and foremost, it is crucial to make sure that a dependable and expandable hosting infrastructure is available. The computational power needed by this infrastructure should be able to perform the data processing as well as analysis that computer vision algorithms demand. The capacity, adaptability, and ease of maintenance are all benefits of hosting the software on a cloud-based service, such as AWS or Microsoft Azure. The capacity to dynamically assign assets based on demand is offered by cloud platforms, guaranteeing that the programme can successfully handle changing traffic loads.

f) *Test and Improve Your Web Application*: Following deployment, you must extensively analyse your web application in order to make sure it is operating as intended. On the basis of user input or problems that surface during testing, you might also need to tweak your code and design. The process doesn't end with creating an online application for visual mapping in parking lots. Thorough testing and ongoing improvement are crucial for ensuring its efficacy and dependability. While testing is essential for spotting and fixing possible problems, continuing development keeps the programme optimised and in line with user needs. In this final section, discussion of the value of testing and how to make the internet-based application using computer vision modelling in parking better will be done. Testing is an essential component of the application's quality assurance process. By assisting in the discovery of faults, failures, and performance bottlenecks, it helps to ensure that the system works as intended. Testing for units, integration evaluation, and system testing are just a few of the testing approaches that can be used. Maintaining a web-based programme for automated vision modelling in parking lots requires constant refinement. It's crucial to keep up with changes in user needs and technological advancements by implementing new features and improvements. This may entail frequent programme upgrades, the addition of new functionality, or the optimisation of already-existing ones in response to user input and developing industry standards. Developers make sure the app is relevant and successful in addressing the changing needs of the parking environment by keeping it up to date.

Table 2 shows the efficiency of the car parking system.

The usage of OpenCV and Python using SQL and a database along with computer vision mapping in parking spot counting has produced encouraging results. This method has shown to be an effective and economical method for managing parking lots since it can correctly identify and keep track of parked cars in real-time. The capacity to give real-time surveillance and updates on parking availability is one of the main benefits of employing computer vision mapping for parking spot counting. Parking lot managers must physically inspect the parking lot to

Table 2. Efficiency of the car parking system.

Metric	**Value**
Detection accuracy	True positive rate: 0.85 False positive rate: 0.05 Precision: 0.92, Recall: 0.85, F1 score: 0.88
Parking occupancy	Average occupancy: 70%, Peak occupancy: 90%, Lowest occupancy: 50%
Parking availability	Real-time availability: 45 spots available out of 60 total, Heatmap of availability: see Fig. 5
Processing time	Average time per frame: 50ms, Total processing time for 24 hours: 43 minutes
User interface	See Figs. 2; 3

determine the number of free spaces when using conventional methods like manual method or sensors. Using computer vision mapping, however, the system is able to Abdullah et al. [6] continuously scan the parking lot and deliver updates on open spots. The outcomes in accuracy as well have been outstanding. Even in challenging situations like occlusion or overlapping, the system was able to detect and categorise cars effectively because to the use of cutting-edge technologies like Qiu et al. [7] deep learning and CNNs. Compared to conventional procedures, this has led to higher accuracy rates, decreasing the possibility of counting errors. Also, the use of a databases has made it simple to save and retrieve data, enabling analysis and optimization of parking lot management. Additionally, this information can be utilised to determine parking trends, forecast peak times, and improve parking lot layout efficiency. The combination of OpenCV as well as Python with SQL and a database for parking spot counting using computer vision mapping has produced promising results. It has shown to be a reliable, affordable, and successful way to manage parking lots, with the potential ability to completely change the parking business.

5. Conclusion

Hussain et al. [1] parking spaces may be precisely counted using computer vision modelling, which is a difficult challenge for traditional counting methods like manual counting or sensor-based monitoring. Even in complex parking lots, computer vision mapping utilises cutting-edge methods, such as machine learning, to accurately measuring and count parking spaces. Computer vision mapping also makes it feasible to monitor parking spots in real-time, which aids in managing and optimising parking lots. This method is more efficient and affordable because it requires less upkeep and costs than traditional techniques like setting up sensors or manual counting.

Qui et al. [7] submitted that due to varying lighting conditions, shifting car park layouts, and erroneous detections, computer vision mapping in car park counts confronts difficulties. The system needs algorithms that can adjust to changing lighting effects because elements like direct sunshine, shaded locations, and artificial lighting might affect how accurate it is. To reliably identify and count parking spaces, the system also needs to be trained on different parking lot layouts. False detections from non-vehicle items, such as bicycles or people, can also happen, necessitating the employment of sophisticated techniques like object identification and classification. It can also be confirmed that the thing being watched is a vehicle by using type of sensors like infrared or ultrasonic technology.

Abdullah et al. [6] proposed a potential method for managing parking lots that can increase accuracy, simplify operations, and provide a simple, affordable solution is the use of computer vision modelling, OpenCV, Python, SQL, etc. databases. The purpose of the study is to show the effectiveness of this approach and the potential for data analysis as well as forecasting to improve parking lot management. This technology can offer more accuracy and cost-effectiveness with real-time updates. This method will likely get better with time and development.

The research on counting parking spaces using computer vision mapping using OpenCV and Python, along with SQL and database, was quite precise and successful. Deep learning and CNNs are examples of advanced techniques that improved counting accuracy and decreased counting errors. The SQL interface made it simple to save and retrieve data, allowing for analysis and parking lot management optimisation. The process gives real-time updates and a cheap, effective way to manage parking lots. The parking industry may undergo a revolution if this method is further developed and improved.

Wang et al. [21] submitted that Python and OpenCV are excellent choices for creating end-to-end parking place counting solutions because they offer a great level of customisation and system integration options. They are capable of handling real-time applications and have superior performance optimisation. Even in difficult circumstances, machine learning techniques like deep learning as well as CNNs could reliably detect and categorise automobiles. Advanced machine learning capabilities are also available in Python packages like TensorFlow and Keras, which can be readily combined with OpenCV to produce more effective solutions.

References

[1] Hussain, A., Rana, M.S. and Khayam, S.A. 2020. A survey of deep learning techniques for parking lot management. *Journal of Parallel and Distributed Computing*, 144: 101–112. Doi: 10.1016/j.jpdc.2020.07.013.

[2] Abadi, M., Barham, P., Chen, J., Chen, Z., Davis, A., Dean, J. and Ghemawat, S. 2016. TensorFlow: A system for large-scale machine learning. pp. 265–283. *In*: *12th {USENIX} Symposium on Operating Systems Design and Implementation ({OSDI} 16).*

[3] Bradski, G. 2000. The OpenCV Library. *Dr. Dobb's Journal of Software Tools*, 25(11): 120–126.

[4] Cao, J., Liu, X., Zhao, X. and Cao, J. 2017. Vehicle detection and parking space tracking for intelligent car park systems. *IEEE Transactions on Industrial Electronics*, 64(10): 8409–8418.

[5] Chen, W., Zheng, Z., Xia, J. and Wu, Y. 2019. Parking space detection based on YOLOv3 and SSD in outdoor environments. *IOP Conference Series: Earth and Environmental Science*, 271(6): 062026.

[6] Abdullah, H., Rahman, M.M. and Hossain, M.O. 2018. Real-time parking lot vehicle detection system using OpenCV and raspberry Pi. *IEEE Access*, 6: 75414–75424. Doi: 10.1109/ACCESS.2018.2880523.

[7] Qiu, H., Wang, C., Chen, Y. and Wu, J. 2018, May. A real-time parking lot occupancy detection system using deep learning. *IEEE Transactions on Industrial Informatics*, 14(5): 2085–2094. Doi: 10.1109/TII.2017.2787758.

[8] Islam, M.M., Ahamed, T., Rahman, M.A. and Hossain, M.A. 2020. Vehicle detection and counting system in parking areas using computer vision. *International Journal of Computer Science and Network Security*, 20(2): 76–82.

[9] Kim, J., Lee, S. and Lee, S. 2021. A deep reinforcement learning based indoor parking space recognition system. *Sensors*, 21(7): 2279.

[10] Li, W., Zhang, Y., Liu, Y., Yan, S. and Li, X. 2016. A parking space detection method using visual saliency based on convolutional neural network. *Neurocomputing*, 173: 814–821.

[11] Al-Musawi, M.R., Abu Bakar, S.A.S., Al-Mansoori S.A.H. and Al-Fahdawi, A.S. 2019. Intelligent parking system based on image processing techniques and wireless sensor networks. pp. 1–6. *In*: *2019 IEEE 5th International Conference on Engineering Technologies and Applied Sciences (ICETAS)*, Bangkok, Thailand, Doi: 10.1109/ICETAS.2019.8934777.

[12] Nguyen, H.M., Nguyen, T.T., Nguyen, D.H., Nguyen, H.D. and Thai, V.T. 2019. A deep learning approach for parking space detection using convolutional neural network. *IEEE Access*, 7: 103105–103117.

[13] Raji, J.A., Adebiyi, A.A. and Sholarin, M.A. 2020. Intelligent car parking system using computer vision and internet of things. pp. 904–908. *In*: *2020 International Conference on Computing, Networking and Communications (ICNC)*. IEEE.

[14] Shao, Z., Tian, Y. and Zhang, Y. 2019. Enhanced parking space detection based on YOLOv3. pp. 1867–1872. *In*: *2019 IEEE International Conference on Mechatronics and Automation (ICMA)*. IEEE.

[15] Wang, C., Li, J., Li, S. and Yan, F. 2020. Car park detection based on deep learning and machine vision in outdoor settings. *International Journal of Distributed Sensor Networks*, 16(9): 1550147720955127.

[16] Wang, J., Yang, L., Guo, R. and Zhang, Z. 2018. Outdoor parking space detection based on SVM with HOG features. pp. 4952–4957. *In*: *2018 37th Chinese Control Conference (CCC)*. IEEE.

[17] Wu, Y., Lim, J. and Yang, M.H. 2016. Object tracking benchmark. *IEEE Transactions on Pattern Analysis and Machine Intelligence*, 37(9): 1834–1848.

[18] Yuan, Y., Wang, S., Lu, L. and Yang, X. 2019. A smart parking system based on internet of things and cloud computing. pp. 1140–1145. *In*: *2018 IEEE 3rd Advanced Information Technology, Electronic and Automation Control Conference (IAEAC)*, Chongqing, China, Doi: 10.1109/IAEAC.2018.8573424.

[19] Yang, M., Liu, Q. and Liu, Y. 2020. A real-time parking space detection method based on deep learning. pp. 63–67. *In*: *2020 IEEE International Conference on Smart City and Sustainable Computing (ICSCSC)*. IEEE.

[20] Yoo, J., Kang, B. and Song, Y. 2019. A LIDAR-based car park recognition system with three-dimensional mapping. *Sensors*, 19(16): 3511.

[21] Wang, Z., Huang, L. and Zhang, Q. 2016. A multi-camera parking lot monitoring system based on OpenCV and android platform. pp. 392–396. *In*: *2016 International Conference on Computer Science and Applications (CSA)*, Wuhan, China.

[22] Zhang, H., Wang, X. and Cao, Y. 2018. Intelligent parking lot system based on computer vision technology. pp. 298–300. *In*: *2018 IEEE International Conference on Applied System Innovation (ICASI)*. IEEE.

[23] Zhang, W., Shen, J., Zhang, Y. and Yang, X. 2018. Real-time parking space detection and tracking using an improved Faster R-CNN. *IEEE Transactions on Intelligent Transportation Systems*, 19(8): 2574–2584.

[24] Zhang, Y., He, H., Lin, H. and Liu, Y. 2019. Real-time parking occupancy prediction with deep learning. *IEEE Transactions on Intelligent Transportation Systems*, 21(4): 1688–1698.

[25] Zhou, W., Huang, Q., Chen, X., Zhu, Y. and Huang, D. 2018. A deep learning-based car park recognition system. *Journal of Physics: Conference Series*, 1073(4): 042022.

Chapter 4

Innovative, Conceptual, and Analytical Approaches for Cyber Security, IoT and Blockchain

Ramesh Ram Naik,[1,*] *Umesh Bodkhe,*[1] *Rohit Pachlor,*[2] *Sunil Gautum*[1] and *Sanjay Patel*[1]

1. Introduction

The interaction between technology and society has created a complex web of vulnerabilities, threats, and opportunities that must be understood and addressed to ensure the security and well-being of individuals, organisations, and nations.

Cybersecurity research focused on the interaction of technology and society aims to examine the intricate relationship between these two domains. It delves into the ways in which technological advancements influence societal dynamics and how societal factors, in turn, shape

[1] Nirma University, Ahmedabad. Gujarat.
[2] MIT ADT University, Pune Maharashtra.
Emails: gautamsunil.cmri@gmailcom; rohitpachlor@gmail.com
* Corresponding author: rrnaik2686@gmail.com

the development and impact of technology. This interdisciplinary field explores the intricate interplay between technology, human behaviour, ethics, privacy, policy, and more.

The study of technology and society encompasses a wide range of areas, including the social and psychological implications of emerging technologies, the impact of cybersecurity threats on individuals and communities, the influence of social media on public discourse, the ethical considerations surrounding data privacy and surveillance, and the role of policy and regulation in safeguarding against cyber threats.

By understanding the complex dynamics between technology and society, researchers can identify vulnerabilities, anticipate potential risks, and develop effective strategies and solutions to mitigate cybersecurity threats. This research plays a crucial role in informing policymakers, organisations, and individuals about the challenges and opportunities presented by the ever-evolving technological landscape.

The Internet of Things (IoT) has emerged as a revolutionary force in a connected society, revolutionising businesses and changing how we interact with technology. Smart homes, wearable technology, industrial sensors, and autonomous vehicles are just a few examples of the IoT devices that have a significant impact on our daily lives. However, a growing concern for cybersecurity is brought on by the rapid expansion of IoT devices.

The need for stronger cybersecurity has become critical as the market for IoT devices keeps growing. The attack surface for possible cyber threats grows quickly as more gadgets are connected to the internet. IoT devices frequently gather and communicate sensitive data, which makes them desirable targets for cybercriminals looking to take advantage of flaws to obtain unauthorised access or steal sensitive data.

The consequences of inadequate cybersecurity in the IoT realm can be severe. Compromised IoT devices can be used as entry points for larger-scale attacks, enabling cybercriminals to infiltrate networks, disrupt critical infrastructure, and even compromise personal safety. Instances of IoT-related breaches and data leaks have already demonstrated the potential for significant financial, privacy, and security implications.

Recognising the urgency of the situation, governments, organisations, and individuals are increasingly demanding better cybersecurity measures to safeguard IoT devices and the networks they connect to. Efforts are being made to establish industry-wide standards and best practices for IoT security, such as secure device authentication, data encryption, and secure firmware updates. Cybersecurity professionals are actively developing innovative solutions to address the unique challenges posed by the IoT landscape.

When it comes to cybersecurity, IoT, and blockchain, there are several innovative conceptual and analytical approaches that can enhance security measures and address emerging challenges. Here are a few examples:

Threat Intelligence Integration: Integrating threat intelligence from various sources and analysing it in real-time can help identify potential cyber threats and vulnerabilities. By leveraging machine learning and artificial intelligence techniques, organisations can analyse vast amounts of data to proactively detect and respond to cyber-attacks.

Behavior Analytics: Utilising behaviour analytics can help identify anomalous patterns and behaviours in IoT devices and blockchain networks. By establishing a baseline of normal behaviour, any deviations can be flagged as potential security breaches. This approach can be useful in detecting unauthorised access, data exfiltration, or malicious activities.

Blockchain-enabled Security: Blockchain technology itself can be used to enhance security in various domains. In the context of cybersecurity, blockchain can be leveraged to create immutable logs of security events, ensuring transparency and tamper resistance.

Secure IoT Architectures: Designing secure IoT architectures involves considering security from the ground up. Implementing security measures such as encryption, secure bootstrapping, and access controls at every layer of the IoT ecosystem can mitigate potential risks.

Threat Modelling: Adopting a proactive approach, threat modelling involves identifying potential threats, vulnerabilities, and risks associated with IoT and blockchain systems. By systematically analysing the system's architecture, components, and potential attack vectors, organisations can develop effective security controls and countermeasures.

Security Automation and Orchestration: As the number of connected devices and blockchain deployments increase, manual security management becomes impractical. Implementing security automation and orchestration tools can help streamline security operations, improve incident response times, and enable quick remediation of security incidents.

Privacy-Preserving Techniques: Preserving privacy is crucial in IoT and blockchain applications. Techniques such as homomorphic encryption, zero-knowledge proofs, and differential privacy can be applied to protect sensitive data while still allowing secure processing and analysis.

Machine Learning for Anomaly Detection: Machine learning algorithms can be trained to detect anomalies and unusual patterns in network traffic, device behaviour, and blockchain transactions. By continuously monitoring and analysing data, these algorithms can identify potential threats or malicious activities that traditional rule-based systems might miss.

These are just a few innovative approaches within the vast field of cybersecurity, IoT, and blockchain. Advancements in technology continue to drive new conceptual and analytical strategies to stay ahead of emerging threats and secure digital ecosystems effectively.

Cybersecurity, IoT (Internet of Things), and Blockchain are three interconnected areas that can work together to enhance security and privacy in various applications. Here are some examples of how these technologies can be applied.

Secure IoT Communication: Blockchain can provide a decentralised and tamper-proof ledger for storing IoT data, ensuring data integrity and authentication. Additionally, blockchain-based smart contracts can enable secure and automated interactions between IoT devices.

Supply Chain Management: Blockchain technology can be used to enhance transparency, traceability, and security in supply chain management. By recording transactions and data related to the supply chain on a blockchain, it becomes difficult to tamper with or manipulate information.

Identity Management: Blockchain-based identity management systems can provide a secure and decentralised approach to identity verification. Users can have control over their personal information and grant access to specific entities when needed. IoT devices can be integrated into this system to provide additional authentication factors, such as biometrics or location data, to enhance security.

Smart Home Security: IoT devices are commonly used in smart homes, but they can also introduce security vulnerabilities. Blockchain can be employed to secure smart home devices and their interactions. By utilising decentralised consensus mechanisms, unauthorised access and tampering can be prevented. Blockchain can also enable secure sharing of data between different devices within the smart home ecosystem.

Cyber Threat Intelligence: Blockchain can facilitate the sharing of cyber threat intelligence among different organisations while preserving privacy and confidentiality. By using blockchain-based platforms, organisations can securely share threat information, such as indicators of compromise, without revealing sensitive details. This collaborative approach can help detect and mitigate cyber threats more effectively.

Decentralised Cloud Storage: Blockchain-based decentralised storage platforms can provide secure and private storage for sensitive data. IoT devices can encrypt and store data on the blockchain, ensuring that it remains secure and accessible only to authorised parties. This eliminates the need for a centralised cloud storage provider, reducing the risk of data breaches.

These examples illustrate the potential of combining cybersecurity, IoT, and blockchain technologies to create secure and trusted systems in various domains.

2. Literature Survey

The rapid advancements in technology have led to the widespread adoption of IoT devices and blockchain technology, revolutionising various industries. However, these developments have also introduced new cybersecurity challenges. This literature review aims to explore innovative conceptual and analytical approaches in the field of cybersecurity, specifically focusing on IoT and blockchain.

This paper provides an overview of the security challenges in IoT, including device vulnerabilities, privacy concerns, and data integrity issues. It discusses innovative approaches such as anomaly detection, intrusion detection systems, and secure communication protocols [1].

The authors present an extensive survey of IoT security challenges and propose solutions such as access control mechanisms, cryptographic techniques, and security frameworks. They also discuss emerging concepts like blockchain for securing IoT devices [2].

This article explores the potential of blockchain technology to enhance IoT security. It discusses innovative concepts such as blockchain-based access control, tamper-proof logging, and decentralised identity management systems for IoT devices [3].

The authors review various data aggregation techniques for IoT devices and analyse their security vulnerabilities. They propose innovative approaches, including homomorphic encryption, secure aggregation protocols, and blockchain-based solutions to ensure privacy and integrity of aggregated data [4].

This systematic review explores the applications of blockchain in cybersecurity. It covers innovative concepts like blockchain-based threat intelligence, secure data sharing, identity management, and decentralised consensus mechanisms [5].

The authors discuss the potential of blockchain technology for mitigating cybersecurity threats. They explore innovative approaches such as blockchain-based authentication, secure data storage, secure code execution, and tamper-proof logging systems [6].

Blockchain technology [7, 8] facilitates trusted transactions between untrusted network participants as a distributed ledger based on cryptography. Since the original Bitcoin blockchain was introduced in 2008 [9], other blockchain systems, including Ethereum [10, 11] and Hyperledger Outside of the current fiat currencies and electronic voucher systems, fabric [12] have arisen with public and private accessibility. Due to its distinct trust and security features, blockchain technology has recently been the focus of a growing number of scientific studies [13–16] and has sparked a great deal of interest among researchers, developers, and business professionals.

3. New Challenges in IoT, Cybersecurity, and Blockchain

The rapid advancement of technology has brought about numerous transformative innovations, and three areas that have gained significant attention in recent years are the IoT, cybersecurity, and blockchain. These domains have revolutionised various sectors, such as industry, healthcare, finance, and transportation, among others. However, as with any technological progress, new challenges have emerged that require careful consideration and innovative solutions to ensure the secure and efficient operation of these systems.

IoT refers to the network of interconnected physical devices embedded with sensors, software, and connectivity capabilities. These devices collect and exchange data, enabling them to interact with their environment and provide valuable insights for businesses and consumers. While IoT has the potential to revolutionise various aspects of our lives, it also poses unique challenges. One significant challenge is ensuring the security and privacy of IoT devices and the data they generate. As more devices become connected, they become potential targets for cyber-attacks, raising concerns about unauthorised access, data breaches, and the compromise of critical infrastructure. Addressing these challenges requires robust security measures, including strong authentication protocols, encryption, and vulnerability management to protect against emerging threats.

Cybersecurity has become a pressing concern in the digital age, as cyber threats continue to evolve in sophistication and scale. With the increasing interconnectedness of devices and systems, cybersecurity vulnerabilities have expanded, creating new challenges for organisations and individuals. Protecting sensitive data, networks, and critical infrastructure from cyber threats is paramount. It demands continuous monitoring, threat intelligence, proactive defence mechanisms, and collaboration between stakeholders to detect, prevent, and respond effectively to cyber incidents. Moreover, as technologies like artificial intelligence (AI) and machine learning (ML) advance, they both present opportunities for enhancing cybersecurity and pose challenges in terms of potential misuse and adversarial attacks.

Blockchain, originally introduced as the underlying technology for cryptocurrencies, has evolved into a versatile tool with applications beyond finance. It is a decentralised and immutable ledger that ensures transparency, trust, and security in transactions and data management. However, the implementation of blockchain technology is not without its challenges. Scalability is a persistent concern, as blockchain networks struggle to handle large volumes of transactions efficiently. Additionally, while blockchain provides inherent security features, vulnerabilities can still arise due to coding flaws, smart contract weaknesses, and human errors. To fully leverage the potential of blockchain, these challenges need to be addressed through innovative consensus mechanisms, interoperability solutions, and rigorous auditing and testing processes.

Indeed, the rapid advancement of technology, particularly in the fields of IoT, cybersecurity, and blockchain, presents several new challenges that require the development of new conceptual and analytical tools. Let's explore some of these challenges.

IoT Security

The IoT connects a vast number of devices, ranging from everyday objects to critical infrastructure systems. This interconnectivity introduces various security concerns, such as data breaches, privacy risks, and vulnerabilities in IoT devices. New conceptual tools are needed to address the unique security requirements of IoT, including secure communication protocols, robust authentication mechanisms, and encryption techniques tailored for resource-constrained devices.

Cybersecurity

As technology evolves, so do the tactics and techniques employed by cybercriminals. Traditional approaches to cybersecurity are no longer sufficient to combat advanced threats. New conceptual tools are necessary to detect, prevent, and respond to sophisticated attacks, such as AI-based threat detection, behaviour analytics, threat intelligence sharing platforms, and proactive security measures like deception technologies.

Blockchain Security

While blockchain technology provides numerous benefits such as decentralised control and immutability, it also presents its own security challenges. Blockchain networks must contend with potential vulnerabilities in consensus algorithms, smart contracts, and key management systems. Developing new analytical tools for auditing blockchain systems, identifying vulnerabilities, and ensuring secure implementation of decentralised applications is crucial.

Privacy and Data Protection

With the proliferation of IoT devices and the increasing digitisation of personal information, protecting user privacy and data becomes paramount. Concepts like privacy-by-design, differential privacy, and secure data sharing frameworks need to be developed and integrated into the design and operation of IoT systems and blockchain networks.

Regulatory and Legal Frameworks

As technology advances, legal and regulatory frameworks must keep pace to address emerging challenges. Developing new analytical tools and frameworks that encompass legal and ethical considerations related to IoT, cybersecurity, and blockchain is vital. This includes aspects such as data governance, liability, jurisdiction, and international cooperation.

Human Factors and User Awareness

Technology security is not solely reliant on technical solutions; it also involves human factors. Enhancing user awareness, education, and promoting best practices are critical for mitigating security risks. Developing new conceptual tools to improve user interfaces, authentication methods, and training programmes can empower individuals to make informed security decisions.

Cybersecurity plays a vital role in protecting both businesses and individuals when it comes to storing, accessing, and retrieving crucial information. It helps safeguard data from unauthorised access, theft, or damage, ensuring confidentiality, integrity, and availability of digital assets.

4. Some Key Reasons Why Businesses Should Prioritize Cyber-Security

Data Protection Cybersecurity measures such as encryption, access controls, and secure data storage systems help protect sensitive information from being compromised. This is particularly crucial for businesses that handle personal, financial, or proprietary data.

Preventing Unauthorised Access

Cybersecurity practices, including strong passwords, multifactor authentication, and network security measures like firewalls help prevent unauthorised individuals from gaining access to sensitive data and systems.

Mitigating Financial Losses

Cyber-attacks can result in significant financial losses for businesses, including costs associated with data breaches, legal liabilities, customer compensation, and damage to the organisation's reputation. Implementing robust cybersecurity measures reduces the risk of such incidents and their associated financial consequences.

Maintaining Business Continuity

Cybersecurity helps ensure that business operations continue uninterrupted. Implementing measures such as regular data backups, disaster recovery plans, and incident response procedures minimise downtime in the event of a cyber-attack or system breach.

Protecting Customer Trust

By prioritising cybersecurity, businesses demonstrate their commitment to protecting customer data and privacy. This builds trust with clients and customers, enhancing the organisation's reputation and competitiveness.

Individuals should also be mindful of cybersecurity best practices to protect their personal information, such as:

Using Strong Passwords: Create unique, complex passwords for each online account and consider using password managers to securely store them.

Enabling Two-Factor Authentication (2FA): Use 2FA whenever available to add an extra layer of security to online accounts. This typically involves providing a secondary verification method, such as a code sent to a mobile device, in addition to a password.

Being Cautious of Phishing Attacks: Be vigilant for suspicious emails, messages, or links that may be attempts to steal personal information. Avoid clicking on unfamiliar links or providing sensitive data without verifying the source.

Keeping Software Updated: Regularly update software, including operating systems, web browsers, and security applications, to ensure you have the latest security patches and protections against known vulnerabilities.

Using Secure Networks: When accessing sensitive information or conducting financial transactions online, use secure and trusted networks. Avoid using public Wi-Fi networks for sensitive activities unless you are utilising a secure VPN (virtual private network) connection.

By following these practices, both businesses and individuals can significantly enhance their cybersecurity posture and protect valuable information from various cyber threats.

Artificial Intelligence and Machine Learning

Conceptual Approach: Leveraging AI and ML algorithms to analyse large volumes of data and identify patterns, anomalies, and potential threats in real-time.

Analytical Approach: Building predictive models that can detect and mitigate cybersecurity risks, identify vulnerabilities in IoT devices, and validate blockchain transactions.

Advantages: AI/ML can process vast amounts of data quickly, adapt to evolving threats, and automate security tasks. They can detect complex attacks and enhance anomaly detection.

Disadvantages: AI/ML models can be prone to adversarial attacks, require substantial computational resources, and may generate false positives/negatives if not properly trained or updated.

Threat Intelligence and Information Sharing

Conceptual Approach: Establishing a collaborative ecosystem where organisations and stakeholders share threat intelligence, vulnerabilities, and attack trends.

Analytical Approach: Analysing shared information to identify emerging threats, patterns, and indicators of compromise (IoCs) to proactively mitigate risks.

Advantages: Enables faster response to new threats, improved situational awareness, and better coordination among organisations. It helps in staying updated about the latest attack vectors.

Disadvantages: Organisations may be reluctant to share sensitive data due to privacy concerns or competitive reasons. Information overload can also be a challenge, requiring efficient data processing and analysis.

Blockchain for Security and Trust

Conceptual Approach: Utilising blockchain's decentralised and tamper-resistant nature to enhance security, transparency, and trust in various applications.

Analytical Approach: Verifying and validating transactions, establishing identity management, securing IoT devices, and ensuring data integrity through distributed consensus mechanisms.

Advantages: Blockchain can prevent tampering, provide immutable records, enhance data integrity, and enable secure peer-to-peer transactions without intermediaries.

Disadvantages: Blockchain has scalability limitations, high computational requirements, and potential governance challenges. Public blockchains can also be vulnerable to 51% attacks.

Zero Trust Architecture

Conceptual Approach: Moving away from perimeter-based security models and adopting a 'trust-no-one' approach that verifies every access request.

Analytical Approach: Implementing multifactor authentication, continuous monitoring, micro-segmentation, and real-time threat detection to validate and authorise each access attempt.

Advantages: Zero Trust provides granular control, reduces the attack surface, and enables better protection against insider threats, lateral movement, and unauthorised access.

Disadvantages: Implementing Zero Trust architecture can be complex and resource-intensive. Legacy systems and dependencies may require significant changes and integration efforts.

Privacy-Preserving Techniques

Conceptual Approach: Implementing cryptographic techniques such as homomorphic encryption, secure multi-party computation, and differential privacy to protect sensitive data.

Analytical Approach: Ensuring data confidentiality, privacy, and compliance while performing analytics, sharing data, or conducting transactions.

Advantages: Privacy-preserving techniques enable secure data processing and sharing without exposing sensitive information, maintaining individual privacy rights.

Disadvantages: These techniques can introduce computational overhead, impact system performance, and require careful implementation to avoid vulnerabilities or cryptographic weaknesses.

Each of these approaches can bring significant benefits to cybersecurity, IoT, and blockchain. However, it's important to consider the specific context, requirements, and limitations of each approach before implementing them in real-world applications. Assessing the trade-offs, conducting thorough risk assessments, and adapting the approaches to the specific use cases are crucial steps to achieve effective and resilient security solutions.

Blockchain technology has gained popularity for its ability to secure data transfers and transactions without the need for intermediaries. While blockchain can provide a secure framework for data transfer, it's important to understand how it achieves this.

Blockchain is a decentralised and distributed ledger technology. It consists of a chain of blocks, where each block contains a list of transactions. These blocks are linked together using cryptographic hashes, creating an immutable and transparent record of all transactions.

Here's how blockchain can facilitate secure data transfer

Decentralisation: Blockchain operates on a peer-to-peer network where multiple participants, known as nodes, maintain and validate the blockchain. This decentralised nature ensures that no single entity has complete control over the data, reducing the risk of hostile parties intercepting or tampering with the information.

Consensus mechanism: Blockchain networks use consensus mechanisms like Proof-of-Work (PoW) or Proof-of-Stake (PoS) to agree on the validity of transactions. These mechanisms require nodes to reach a consensus before a transaction can be added to the blockchain. By ensuring agreement among a majority of nodes, blockchain technology prevents unauthorised changes or malicious attacks on the data.

Cryptographic security: Blockchain employs cryptographic techniques to secure data transfers. Each transaction is digitally signed using the sender's private key, ensuring that only the intended recipient can access and decrypt the data. The use of public-private key pairs and encryption algorithms makes it extremely difficult for hostile parties to intercept or modify the transferred data.

Immutable records: Once a transaction is added to a block and included in the blockchain, it becomes practically impossible to alter. The decentralised nature of the blockchain, coupled with cryptographic hashes linking the blocks, creates a tamper-resistant and transparent record of all transactions. This immutability provides an additional layer of security, as any attempts to modify the data would require significant computational power and consensus from the network.

5. Conclusion

In this chapter, we will explore the field of cybersecurity research focused on the interaction of technology and society. We will delve into key areas of inquiry, highlight notable research findings, and discuss the implications for individuals, organisations, and society as a whole. By examining the multifaceted relationship between technology and society, we can gain valuable insights to shape a secure and resilient future in the digital era. The exponential growth of the IoT device industry has raised concerns about cybersecurity. As more devices become connected, the need for robust security measures becomes increasingly vital. The diverse nature of IoT devices, coupled with resource constraints and long lifecycles, presents unique challenges in ensuring their protection. With the potential risks and consequences associated with IoT-related breaches, the demand for better cybersecurity is growing rapidly, driving the development of enhanced security standards and solutions to safeguard the IoT ecosystem.

Innovative conceptual and analytical approaches play a crucial role in addressing cybersecurity challenges in IoT and blockchain. The reviewed literature emphasises the importance of anomaly detection, secure communication protocols, access control mechanisms, and cryptographic techniques. Furthermore, the integration of blockchain technology shows promise for enhancing security and privacy in both IoT and cybersecurity domains. However, further research and development are needed to fully exploit the potential of these innovative approaches and ensure robust cybersecurity in the evolving technological landscape.

The IoT, cybersecurity, and blockchain domains offer immense possibilities for innovation and disruption. However, they also bring forth new challenges that require constant attention and innovation. By investing in robust security measures, fostering collaboration between stakeholders, and promoting research and development, we can navigate these challenges and unlock the full potential of these technologies for a secure and connected future. The ever-evolving landscape of IoT, cybersecurity, and blockchain necessitates the development of new conceptual and analytical tools. These tools should address the unique security challenges of interconnected devices, advanced cyber threats, blockchain vulnerabilities, privacy concerns, legal considerations, and user awareness. By leveraging innovative approaches, we can enhance the security and resilience of these technologies.

It's important to note that while blockchain technology offers enhanced security for data transfers, it is not a solution for all scenarios. Factors such as the design and implementation of the blockchain network, the security measures of the devices involved, and the encryption techniques used for data transmission can still impact the overall security. Therefore, it is crucial to consider these factors and adopt appropriate security practices when leveraging blockchain for secure data transfer.

References

[1] Abomhara, Mohamed and Geir M. Køien. 2014. Security and privacy in the Internet of Things: Current status and open issues. *2014 International Conference on Privacy and Security in Mobile Systems (PRISMS)*. IEEE.

[2] HaddadPajouh, H., Dehghantanha, A., Parizi, R.M., Aledhari, M. and Karimipour, H. 2021. A survey on internet of things security: Requirements, challenges, and solutions. *Internet of Things*, 14: 100129.

[3] Yu, Y., Li, Y., Tian, J. and Liu, J. 2018. Blockchain-based solutions to security and privacy issues in the internet of things. *IEEE Wireless Communications*, 25(6): 12–18.

[4] Pourghebleh, Behrouz and Nima Jafari Navimipour. 2017. Data aggregation mechanisms in the Internet of things: A systematic review of the literature and recommendations for future research. *Journal of Network and Computer Applications* 97: 23–34.

[5] Le, Tuan-Vinh and Chien-Lung Hsu. 2021. A systematic literature review of blockchain technology: Security properties, applications and challenges. *Journal of Internet Technology* 22.4: 789–802.

[6] Gimenez-Aguilar, M., De Fuentes, J.M., Gonzalez-Manzano, L. and Arroyo, D. 2021. Achieving cybersecurity in blockchain-based systems: A survey. *Future Generation Computer Systems*, 124: 91–118.

[7] Aste, Tomaso, Paolo Tasca and Tiziana Di Matteo. 2017. Blockchain technologies: The foreseeable impact on society and industry. *Computer* 50.9: 18–28.

[8] Zheng, Z., Xie, S., Dai, H., Chen, X. and Wang, H. 2017, June. An overview of blockchain technology: Architecture, consensus, and future trends. pp. 557–564. *In*: *2017 IEEE International Congress on Big Data (BigData Congress)*.

[9] Nakamoto, Satoshi. 2008. Bitcoin: A peer-to-peer electronic cash system. *Decentralized Business Review*.

[10] Wood, Gavin. 2014. Ethereum: A secure decentralised generalised transaction ledger. *Ethereum Project Yellow Paper* 151.2014: 1–32.

[11] Buterin, V. 2014. A Next-Generation Smart Contract and Decentralized Application Platform, Etherum [Online]. Available: http://buyxpr.com/build/pdfs/EthereumWhitePaper.pdf.

[12] Androulaki, E., Barger, A., Bortnikov, V., Cachin, C., Christidis, K., De Caro, A. and Yellick, J. 2018, April. Hyperledger fabric: a distributed operating system for permissioned blockchains. pp. 1–15. *In*: *Proceedings of the Thirteenth EuroSys Conference*.

[13] Kan, L., Wei, Y., Muhammad, A.H., Siyuan, W., Gao, L.C. and Kai, H. 2018, July. A multiple blockchains architecture on inter-blockchain communication. pp. 139–145. *In*: *2018 IEEE International Conference On Software Quality, Reliability and Security Companion (QRS-C)*.

[14] Miller, D. 2018. Blockchain and the internet of things in the industrial sector. *IT Professional*, 20(3): 15–18.

[15] Fiaidhi, J., Mohammed, S. and Mohammed, S. 2018. EDI with blockchain as an enabler for extreme automation. *IT Professional*, 20(4): 66–72.

[16] Samaniego, M., Jamsrandorj, U. and Deters, R. 2016, December. Blockchain as a Service for IoT. pp. 433–436. *In*: *2016 IEEE International Conference on Internet of Things (iThings) and IEEE Green Computing and Communications (GreenCom) and IEEE Cyber, Physical and Social Computing (CPSCom) and IEEE Smart Data (SmartData)*.

Chapter 5

Mitigating Security Vulnerabilities in the Internet of Things

An Examination of Blockchain-based Solutions

Ansh Suresh Bhimani,[1] *Rohan Rakeshkumar Shah*[2] and *Kaushal Arvindbhai Shah*[3,*]

1. Introduction

Today we are living in a digital era, where the Internet of Things (IoT) has changed our way of living. But every coin has two sides, so does this. Despite having numerous benefits, the concerns regarding security and privacy in the existing IoT infrastructure pose a significant challenge to its expansion. Recent data breaches and privacy violations in IoT

[1] G-202, Palm Residency, Near Shukan Cross Road, Nikol, Ahmedabad, Gujarat, India-382350.

[2] H-13, Shikhar Apartments, Nr. Cadila Bridge, Opp. Jiviba School, Ghodasar, Ahmedabad, Gujarat, India-380050.

[3] Pandit Deendayal Energy University, Gandhinagar, Gujarat, India.
Email: ansh.bhimani@outlook.com; rohanshah120603@gmail.com

* Corresponding author: shah.kaushal.a@gmail.com

have led to user insecurity and hindered the technology's growth. To address these concerns, this chapter explores the use cases of Blockchain technology to provide a secure and transparent infrastructure for IoT.

Vision

This chapter envisions helping accommodate Blockchain Principles, i.e., transparency, reliability, and traceability to facilitate a trustworthy environment for the exchange of information between IoT devices connected to the network [1–3]. This is very important keeping in mind the sensitivity of the data that is shared across devices and to promote a widespread adaptation of IoT technology and pave the way for further innovation.

Objective

The objective of this chapter is twofold. First, to examine the potential of emerging technologies, such as blockchain, to solve security issues in IoT, and second, to implement blockchain in Industrial IoT.

Contribution

The contribution of this chapter is threefold:

- Identifying security constraints in the existing IoT infrastructure
- Proposing a blockchain-based framework that addresses these vulnerabilities
- Analysing the proposed framework by comparing it with existing frameworks

A blockchain-based framework would be proposed which would address the limitations of existing infrastructure and try to integrate blockchain and IoT by considering the challenges faced in maintaining privacy and security in the existing infrastructure [4, 5]. The framework would focus on using the core concepts of blockchain technology, such as decentralisation and consensus mechanisms, to facilitate a secure and transparent environment for exchanging information between IoT devices.

A blockchain-based framework would be proposed targeting to eradicate the challenges of the existing infrastructure in maintaining privacy and security. This framework will use the core blockchain concepts like decentralisation and consensus mechanisms to architect a platform aimed at felicitating the exchange of information between IoT devices.

A head-on comparison between the classical solutions for IoT security with the proposed blockchain-based solution will be discussed at the end of this chapter.

Conclusion

In conclusion, this chapter seeks to establish a secure, transparent, and traceable infrastructure for IoT that will help build trust among nodes in the network and promote the widespread adoption of this exciting and rapidly expanding technology. The proposed blockchain-based framework offers a novel solution to IoT security and privacy concerns and significantly contributes to the field. By combining the strengths of blockchain and IoT, this chapter seeks to provide a secure and trustworthy infrastructure for exchanging information, promoting the growth and adoption of IoT technology.

Related Works

Numerous issues still need to be resolved even though much research is being conducted on fusing blockchain with IoT, with some offering different designs and solutions. These include the IoT device's shortcomings in maintaining privacy and security [6, 7]. Despite the attempts to develop an effective and efficient architecture, there are still numerous limits, necessitating a design that considers these difficulties.

2. Introduction to IoT

The Internet of Things (IoT) is a cluster of interconnected devices, machines, and objects, equipped with various technologies such as sensors, and software, which are connected over the internet, allowing for the collection and analysis of data, and are embedded with unique identifiers as well as the ability to transfer data over a network without requiring human-to-human or human-to-computer interaction.

2.1 IoT Communication and Architecture

2.1.1 Sensing Layer

The Sensing layer collects data from the physical layers, which consist of sensors and actuators. This layer captures real-world conditions as different data sets like temperature, humidity, air pressure, wind speed, etc. The sensors used here could also be embedded into different microcontrollers to collect data from various types of machinery in industries. This information is then carried over by the network layer, where analysis and calculations would be performed to improve productivity and better decision-making.

2.1.2 Network Layer

The Network layer is a critical aspect of the whole IoT architecture as it looks after the communication among the devices and between the device and the gateway through wired or wireless networks. This layer looks after the network protocols incorporated for optimal data transmission and security. The network typically would include a gateway that would act as an entrance for the inflow of network traffic. It would enforce the protocols and route the traffic to its desired location [8].

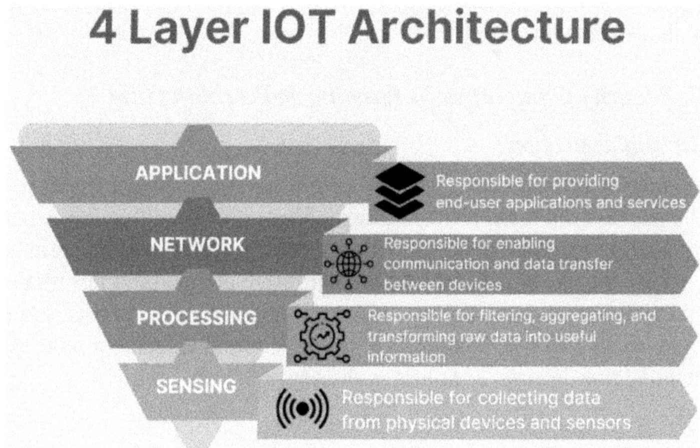

Fig. 1. IoT 4-layered architecture.

2.1.3 Data Processing Layer

The Data Processing layer processes and analyses the data collected from the IoT devices to extract relevant information. This analysis includes the filtration of data from datasets and its segregation. This layer contains Cloud computing systems, which provide robust data processing and storage capability for large-scale data analysis. Hence, this layer plays a crucial role in extracting valuable insights from the data collected and making the most out of it.

2.1.4 Application Layer

The Application layer in IoT architecture is the innermost level and is responsible for providing the end user with the desired functionality and services. This layer interacts with the underlying layers to extract meaningful insights from the collected data, make decisions based on that data, and perform the necessary actions. It acts as a bridge between the IoT system and the user, providing an interface to monitor and control the system. This layer is implemented using APIs (application programming interface). The main goal of this Layer is to provide an intuitive, user-friendly interface to interact with the IoT system and to present the processed data to the end user in a meaningful and valuable way. This layer is also responsible for the IoT system's security, privacy, and data management [9].

2.2 Security Challenges in Existing IoT Architecture

2.2.1 Authentication

Methods for authenticating network communication between nodes need to be improved in the current paradigm. There is no specified mechanism for ensuring that the data received by another node is valid and reliable. There are very few tested methods to ensure that incoming information is reliable but are not efficient. Hence, it has given rise to trust issues. It may occur that the nodes do not reach a mutual agreement with each other and thus leading to failure in the network [10].

2.2.2 Traceability

When a series of events take place on the network, there is no way to trace back in time and review a particular event that has occurred. This problem in existing infrastructure would lead to trust issues among the nodes as there would be no accountability in the network.

For example, if in a network, some IoT device has malfunctioned, and we want to know what the reasons are for it or what was the last communication that it had with other devices, then it is not possible with the existing architecture [11].

2.2.3 Single Point of Failure

A central hub/router is an essential requirement to facilitate communication in an IoT network. If the hub fails, the communication will be cut off, leading to the total shutdown of the network, eventually making the network vulnerable to Distributed Denial of Service (DDoS) attacks [12].

Also, if in a centralised system, a server fails, then all the data stored in it would be gone. There would be no way to retrieve those data. In case a central server is used to store and authenticate the data, that server may be highjacked, and taken over by someone else, leading to disruption in the network.

2.2.4 Trust and Privacy

Today the increasing number of nodes in an IoT network has given rise to trust issues. Ensuring that all the data transmitted over the network is trustable and reliable is challenging. Also, if the number of nodes has increased, handling the access rights to the resources would take much work. More transparency is expected in the existing infrastructure, where all communication is routed through a centralised hub, where chances of DDoS attacks prevail. The nodes may need to learn about what happens with the data transmitted by them.

2.2.5 Distributed Denial of Service Attacks

The DDoS attack on a network makes the resources unavailable to intended users due to an overload of requests flooding the host [13]. The attackers exploit vulnerabilities in the system and take control of multiple nodes in the network, making it difficult to control the attack.

2.3 Classical Solutions to IoT Security Challenges

2.3.1 Modern Methods of Authentication

Modern methods of authentication could be used like biometrics, face recognition, digital certification, and multifactor authentication. This would help in reducing the chances of security breaches.

2.3.2 Confidentiality

In the context of the IoT, it is crucial to ensure data security and privacy by implementing encryption mechanisms to protect exchanged data from malicious attacks. Cryptographic solutions can ensure data confidentiality, but many existing algorithms are computationally and storage-intensive, making them inefficient or impractical for IoT devices with limited resources. To address this issue, researchers have proposed various cryptographic solutions that are optimised for resource constrained IoT devices [14]. These solutions fall into two main categories: symmetric and asymmetric cryptographic solutions.

2.3.3 Availability

The security of IoT systems requires the protection of the availability of the system from malicious attacks and unintentional failures. This is particularly important because violations of availability can lead to significant economic losses or safety damages. Attackers can exploit vulnerabilities at different levels of the system, such as network, software design, or cryptographic algorithms.

2.3.4 Traditional Public Key Solutions

The traditional way of using cryptography involves using public keys and issuing certificates to users. There are different methods like RSA (Rivest-Shamir-Adleman), DSA (Digital Signature Algorithm), and others, but they require a lot of energy and are not good for small devices. However, some methods like NTRU (N-Th degree Truncated Polynomial Ring Cryptosystem) and elliptic curves are less energy-intensive and provide good security. Two papers proposed methods for secure communication in IoT devices while protecting user privacy. One method, called DQAC (Dynamic Quorum-Based Access Control), ensures authentication and confidentiality, while the other uses ring signatures to grant access without revealing users' identities. Both methods were tested and shown to work well in real IoT applications.

3. Introduction to Blockchain

Blockchain is a distributed, immutable ledger that facilitates the process of recording transactions, that are securely linked together with the help of cryptographic hash functions, where each transaction is stored in a form of a block (a nomenclature for data nodes) which stores cryptographic hash of the previous block, a cryptographic hash of this block, a timestamp, and transaction data.

3.1 Blockchain Architecture

The blockchain is made of different blocks of data linked together (similar linking like linked list data structure) where each block contains some data [15].

The following data is stored on the blocks:

1. *Previous node Hash*

   The node in the network contains a hash code of the previous node, which is useful for tracing to the next hash. This hash is useful in traversing back in the chain of nodes, to retrieve the previous data. This hash of the current node is generated by the SHA-256 algorithm. This is done to create a unique hash.

2. *Current node Hash*

   The hash of the current node is also stored in the block. This hash of the current node is generated by the SHA-256 algorithm. This is done to create a unique hash which converts the private key to a public key.

3. *Timestamp*

   The block is also assigned the date and time (jointly called timestamp) of its creation. This helps us to track down when and who created the block. This helps in the unique identification of the block.

3.2 Types of Blockchain

There are three main types of blockchain: public, private, and consortium.

3.2.1 Public Blockchain

A public blockchain is a decentralised system where anyone can join and participate in the network. These are open to everyone and can be accessed at any time by anyone. Examples of public blockchains include Bitcoin and Ethereum.

3.2.2 Private Blockchain

A private blockchain is a permissioned system where access is restricted to a specific group of users. Private blockchains are often used for internal company operations, where security and privacy are important. Examples of private blockchains include Hyperledger Fabric and Corda.

3.2.3 Consortium Blockchain

A consortium blockchain is a hybrid of public and private blockchains. It is a permissioned system where a group of organisations comes together to maintain the network. Consortium blockchains are often used for industry-specific applications where multiple organisations need to share data securely and transparently [16].

3.3 Consensus Mechanisms

Consensus mechanisms are the rules that determine how a blockchain network reaches an agreement on the validity of transactions and updates to the ledger. Consensus mechanisms are used to ensure that all participants in a blockchain network agree on the current state of the ledger. Different consensus mechanisms have different trade-offs in terms of performance, security, and energy efficiency, and the choice of mechanism will depend on the specific needs of the blockchain application.

Some commonly used consensus mechanisms are [17]:

3.3.1 Proof of Work (PoW)

Proof of Work is a consensus mechanism that is used in several blockchain networks, including Bitcoin and Ethereum. In this mechanism, miners compete to solve a cryptographic puzzle that requires a large amount of computational power. Once the puzzle is solved, the miner can add a new block to the blockchain, and they receive a reward for doing so. PoW is a secure consensus mechanism, but it is also energy-intensive and slow.

3.3.2 Proof of Stake (PoS)

PoS is an alternative consensus mechanism to PoW. In this mechanism, validators are chosen based on the amount of cryptocurrency they hold, and they are given the right to create new blocks. Validators put up a stake in their cryptocurrency as collateral, and if they are caught cheating or maliciously validating, they lose their stake. PoS is more energy efficient than PoW, but it can be more centralised because those who hold the most cryptocurrency have the most power.

3.3.3 Proof of Authority (PoA)

PoA is a consensus mechanism that is mostly used in private blockchain networks. In this mechanism, validators are chosen based on their reputation or identity, rather than their computational power or stake. PoA is more efficient than PoW, PoS, and DPoS, but it is also more centralised.

3.3.4 Byzantine Fault Tolerance (BFT)

BFT is a consensus mechanism that is used in private blockchain networks. In this mechanism, validators are chosen based on their reputation or identity, and they must agree on each transaction for it to be added to the blockchain. BFT is very secure and efficient, but it is also more centralised than other consensus mechanisms.

3.3.5 Practical Byzantine Fault Tolerance (PBFT)

PBFT is an upgraded version of BFT. In this consensus mechanism, it is ensured that the network reaches a consensus even if some nodes are offline or behaving maliciously. In this, if a node is compromised, then a fixed vote is given to it for the consensus.

Once a node receives enough votes to accept the transaction, it sends a message to all other nodes confirming that the transaction has been accepted. This process continues until all nodes in the group have agreed on the transaction.

3.4 Security Aspects of Blockchain Technology

The security traits of the blockchain-based systems are described below:

3.4.1 Immutability

Immutability is the ability to resist any change or interference with existing data intentionally or unintentionally.

Consistency of data in a blockchain network refers to ensuring that every node in the network always has an identical copy of the data.

The blockchain is considered tamper-resistant as the data stored on the blocks cannot be modified or tampered with. There are two cases of tampering, by the miner or by an opponent, but the hash function, digital signatures, and mining via nodes ensure the transactions' security. Any attempts to modify the data would lead to a discrepancy in hash values, making it impossible to tamper with the data without being noticed.

This feature helps the nodes to trust the data available on the blockchain as they cannot be manipulated later once the original owner has published it.

3.4.2 Traceability

One of the key features of blockchain is traceability. Every transaction recorded on the blockchain is transparent and visible to every node in the network. This provides an immutable and auditable history of all transactions on the blockchain. The distributed and decentralised nature of the blockchain network ensures that the data recorded on the blockchain cannot be altered, providing a tamper-proof and transparent system. The traceability aspect of blockchain is especially useful in industries such as supply chain management and food safety, where it is essential to track the movement of goods and ensure the authenticity and quality of the products. By providing a transparent and tamper-proof record of every transaction, blockchain helps to increase accountability, reduce fraud, and improve trust in the system.

3.5 Smart Contracts

Smart Contracts are applications of blockchain, which are self-execute table agreements with predetermined conditions, which could be self-executed on the blockchain when some conditions are met. All the transactions that are done via these smart contracts are stored on the blockchain.

In these smart contracts, we must pay some fees (called gas fees) whenever we want to add anything to the network. These gas fees change as per the demand to perform a transaction on the network.

The advantages of smart contracts are:

3.5.1 Savings

Smart contracts can significantly reduce costs associated with traditional legal contracts, such as fees for lawyers, brokers, and other intermediaries. Smart contracts are self-executing, so they eliminate the need for intermediaries and can automate many manual processes. This can lead to significant cost savings, particularly for industries that rely heavily on contracts.

Smart contracts are self-executable, so there is no need for intermediaries between the two parties involved in any transaction, eventually cutting down costs. These are helpful in places with a great need for some intermediaries, which could be eliminated using smart contracts.

3.5.2 Security

Smart contracts are secured by the blockchain, making them immutable and resistant to tampering. They are transparent, and once deployed on the blockchain, they cannot be modified or deleted. This makes smart contracts an ideal solution for industries that require secure and auditable transactions, such as finance and supply chain management.

3.5.3 Confidence and Openness

The use of smart contracts can provide greater transparency and accountability in business transactions. Smart contracts are designed to be open, and their code is publicly available on the blockchain, which can help to build trust between parties. The use of smart contracts can increase confidence in business relationships, particularly in situations where there may be a lack of trust between parties.

3.5.4 Accuracy, Efficiency, and Rapidity

Smart contracts are automated, which means they can execute transactions with accuracy and efficiency. They are designed to be error-free, and once deployed on the blockchain, they can execute without the need for human intervention. Smart contracts can also execute transactions rapidly, leading to quicker settlement times and faster processing of contracts.

Smart Contract Example

```
// SPDX-License-Identifier:MIT
pragma solidity ^0.8.8;
contract Storage
{
 uint256 mobileNumber;

 mapping(string => uint256) nameToMobileNumber;

 struct People
 {
   uint256 mobileNumber;
   string name;
 }

 People[] private people;

 function addPerson(string memory _name, uint256 _mobileNumber)
 public
 {
   People memory newPerson = People(_mobileNumber,_name);
   people.push(newPerson);
   nameToMobileNumber[_name] = _mobileNumber;
 }
 The function retrieves (string memory name) public view
 returns(uint256)
 {
   return nameToMobileNumber[name];
 }
}
```

The code is a simple contract that stores mobile numbers and names of people on the blockchain. However, storing data incurs a cost, while retrieving data is free since it only involves reading the nodes. To store two values (mobile number and name), a structure is used, and the data is stored in an array using the addPerson function, which creates a block on the blockchain. To relate a name with a mobile number, a mapping is used, which helps access the data later. To retrieve the mobile number, the retrieve function takes a name as input and searches for the block containing the data to return it.

4. Blockchain for IoT Security

4.1 Integration of IoT with Blockchain

Integration of current IoT infrastructure with Blockchain technology would strengthen its security. This could be done using the core features of blockchain, i.e., immutability, transparency, etc., to address the issues that we face in the existing infrastructure.

This could be done by introduction of a Blockchain layer in the existing infrastructure where the IoT devices would be communicating with each other. Using blockchain we would store the interactions of the individual IoT device with the network into the Blockchain database. By doing so it is ensured that the network is accountable and the host knows what communication was developed by the nodes with the network.

The abovementioned framework is discussed in detail in the later section of the chapter. Now let's discuss how blockchain helps to mitigate the existing security threats in current IoT infrastructure.

4.2 Tackling Security Threats in IoT Using Blockchain

4.2.1 Authentication

With the help of core Blockchain principles like decentralization and immutability, a systematic-procedure could be formed to authenticate the incoming information and know its genuineness. This could be done by storing the previously known information to predict the new information.

The authentication of transactions in a blockchain network using the PBFT consensus algorithm involves a multiphase process. A client node initiates a transaction request, which is then validated by all nodes in the network through a series of messages.

Once the transaction is validated by a quorum of nodes, it is executed and added to the blockchain. The client node then receives a confirmation message. This method ensures that all nodes in the network agree on the same set of transactions, making the network fault-tolerant and resistant to attacks by malicious actors.

4.2.2 Single Point of Failure

In a traditional IOT network, all the nodes are connected to the central hub which controls the flow of communication between the nodes [18]. If the hub goes down due to some failure, it will lead to the downfall of the whole network, leading to serious damage.

The above situation could be avoided if we use a blockchain-based architecture where all the nodes would be connected on the network and each node could have its copy of the blockchain network. This way if one of the nodes goes down then also the network would not fall and continue to work.

As soon as the node comes online again it could catch up with the other nodes by checking the blockchain's latest node hash and updating the same on its copy which would minimize the problem of a Single Point of Failure.

4.2.3 Trust

Trust in the network could be increased by using blockchain in the current IoT network. Blockchain's property to be transparent and traceable would help in achieving it.

Blockchain's principle of immutability and traceability would help ensure that the data in the blockchain is trustworthy and tamper-resistant. Thus, the other nodes could rely on the data previously available and consider that it was updated by the stated user only.

The data is also updated as per the consensus algorithm which helps in validating the transaction and reaching a common state on the blockchain.

Smart contracts could also be used which would be executed automatically when some condition is triggered and would perform some tasks thus limiting human intervention.

4.2.4 Man-in-the-Middle Attack

Man-in-the-Middle attacks is a significant threat in IoT networks. But these could be mitigated using Blockchain technology. This could be done by using the dual key cryptography feature of blockchain—a public key and a private key.

In blockchain, whenever a node is created it is allotted a private key and by performing a hash function (SHA-256) on it a public key is generated. The private key is used to access the network's data by the node and the public key acts as a unique identification to identify the node.

These keys would help us to identify each of the nodes of the network uniquely and eliminate the risk of man-in-the-middle attacks on the network as we would know whom the node is in the network, without exposing their data.

4.2.5 DDoS Attack

DDoS attacks are attacks where an attacker sends heavy traffic on the network, exceeding the maximum threshold the network can bear leading to the shutting down of the whole network leading to loss of data and time.

To ensure the availability of the blockchain network, decentralised peer-to-peer connections are used, which allow the network to process transactions even when some nodes fail. For the attacker to fully block the blockchain, they would have to compromise more than one-third of the network nodes, which is impractical, in the case of a large number of nodes.

In case of the lower number of nodes, a private blockchain network could be created where access would be granted as per the nodes' wish.

4.3 Proposing Blockchain-based Industrial IoT Framework

Here now we propose a Blockchain-based industrial IoT model to conquest the previously stated issues of the existing IoT model. The model would aim to tackle all the issues that the current infrastructure of IoT faces. The model would have the same layers as the current infrastructure of the IoT model but have additional concepts of blockchain by integrating a private blockchain network.

We are using a private blockchain network as using a public blockchain would result in a breach of data privacy as anyone could enter a public blockchain and view the transactions that are being carried out on the blockchain, while on a private network, only authorised nodes could view the transactions.

The network would run on PBFT consensus mechanism. We would use this consensus algorithm as it would also give us some consensus even if some nodes were not online or behaving in a faulty manner. If on the network more than one-third of the nodes are not available, then no result would come, and the network would be stagnant [19]. This is essential because if more than one-third of members are not available to vote then the network may centralise, and some nodes would gain control of the whole network.

The exact specification is as follows:

1. *Device Layer*

   The device layer would be modified now to identify the incoming requests from the nodes of the network and handle them. It would have an identity management system that would categorise each request from the nodes and help to decide which requests would be granted.

   A new node request is considered a transaction and its information would be stored in new blocks.

   This layer would contain basic information about all the IoT devices like their names, device IDs, and protocols to decide whether to grant a particular request or not.

2. *Network Layer*

   In the proposed architecture, the network layer connects IoT devices through a router, which in turn connects to a data processing layer responsible for handling node requests. This layer ensures end-to-end encryption and avoids single points of failure through node interconnection. Multiple entry points to the blockchain are provided, allowing access via authenticated hubs over the internet. A private blockchain restricts access to authorised nodes only.

   When a node needs to access resources, it sends a request to the router. The router determines access based on smart contract protocols, and the transaction result, along with additional details like the requesting IoT device's ID, timestamp, request status, allocated resources, and usage duration, is stored on the blockchain. This information facilitates auditing of past and future events.

3. *Data Processing Layer*

   The Data Processing Layer processes and analyses the data collected from the IoT devices to extract relevant information. This analysis includes the filtration of data from datasets and its segregation. The data from the IoT devices would be collected from the Network layer and then analytics would be run on it to extract relevant information. This information would generally include the information about the IoT devices and the services that were used by them, time period for which they were used, the active time of the IoT device, resources used, changes it made to the previous state of the network. An off-chain data storage could also be included in this layer to store the data for analysis purpose.

4. *Blockchain Service Layer*

   This layer is responsible for storing the crucial data which is extracted by the Data Processing Layer to be stored on the blockchain. The data stored here would be tamper-proof which would ensure that all the critical communication in the network is stored in form of a ledger which would help us to review critical security events.

   A smart contract would be defined in this layer which would determine how the data that is extracted in the data processing layer would be stored on the blockchain. This smart contract would determine which data would be immutable and which data could be accessed by the users.

5. *Application Layer*

   This layer provides APIs for the user to interact with the architecture. This layer would have pre-defined smart contracts to allow the users to communicate with the IoT devices and perform the available functionalities. It is also possible in this layer to add new functionalities by creating new smart contracts and decentralised applications (DApps) [20].

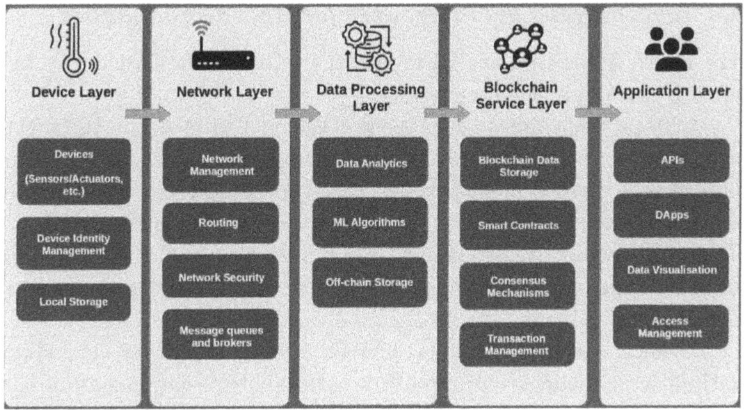

Fig. 2. Layered architecture of the proposed blockchain-based IoT framework.

Table 1. Diverse sector applications of IoT-blockchain framework.

Area of Application	**Usage**
Supply Chain	Transparency, Traceability, Accountability, Security, Efficiency
Healthcare	Secure Data Sharing, Interoperability, Patient-Centricity, Clinical Trials, Drug Supply Chain Management
Energy	Peer-to-peer Energy Trading, Renewable Energy Certificates, Microgrids, Grid Management, Carbon Emissions Trading
Identity Management	Digital Identity Verification, Fraud Reduction, Privacy, Decentralised Identity
Finance	Cross-border Payments, Remittances, Digital Assets, Smart Contracts, KYC/AML
Government	Voting Systems, Public Record Management, Taxation, Land Title Registration, Public Service Delivery
Manufacturing	Inventory Management, Tracking and Tracing, Anti-counterfeiting, Supply Chain Management, Quality Assurance
Agriculture	Livestock Management, Food Supply Chain Traceability, Precision Farming
Autonomous Vehicles	Fleet Management, Smart Parking, Supply Chain Management

Below are some examples of how the proposed IoT-blockchain framework can be applied across various industries:

4.4 Implementing the Framework for Healthcare Industry

The above framework can be implemented for healthcare industry for the use-case of clinical drug testing.

Clinical trials are investigations conducted to evaluate the safety and effectiveness of medical, surgical, or behavioural interventions in human subjects. These studies are crucial for researchers to determine the viability of new treatments, preventive measures, or medical devices. The primary objective of a clinical trial is to assess whether a novel form of treatment surpasses existing methods in terms of efficacy or presents fewer adverse effects.

Clinical research aims to achieve several objectives, including early disease diagnosis, preventive measures, improving quality of life for patients with serious or chronic conditions, and studying the role of caregivers and support groups.

This framework will also be like the above proposed framework and contain five layers:

4.4.1 Device Layer

The device layer would be comprised of various types of IoT devices used in drug clinical trials, such as electrocardiograms (ECGs), blood pressure monitors, and biosensors. These devices can include wearables, sensors, monitoring devices, and data collection tools. Laboratory equipment like analysers, centrifuges, and chromatography machines can also be incorporated with IoT capabilities to automate data collection, improve accuracy, and seamlessly integrate with the overall drug testing framework.

To ensure secure data transmission, each device would be assigned a unique identifier known as a device ID, and associated protocols will be employed.

4.4.2 Network Layer

A router serves as a central gateway connecting IoT devices, facilitating secure and uninterrupted connectivity between the devices, the data processing layer, and the blockchain service layer.

To address the potential single point of failure at the router, an Automatic Failover System is implemented, along with network monitoring and alerting. The system ensures a smooth transition to a backup router in the event of a router failure. Simultaneously, alerts are generated to notify the network administrator of the failure, aiding in issue diagnosis. Routing tables are updated to redirect traffic to the backup router, restoring the system's functionality.

The backup router possesses distinct security configurations, such as a different static IP address, updated DNS (domain name system) servers, and firewall rules. It is also equipped with the necessary configurations and protocols to seamlessly assume network responsibilities from the primary router.

The monitoring system analyses the primary router's network, identifying any suspicious activity and facilitating troubleshooting. Once the fault is identified and resolved, the network is promptly transferred back to the primary router to prevent disruptions to individual nodes caused by new configurations.

By integrating the automatic failover system and network monitoring and alerting, the proposed framework effectively mitigates the single point of failure associated with the router. The failover system ensures uninterrupted network connectivity and services by seamlessly transitioning to the backup router. Simultaneously, the monitoring and alerting system provides real-time visibility, enabling timely detection and resolution of any router-related issues.

4.4.3 Data Processing Layer

In the data processing layer, we would perform analysis on the clinical measurements such as blood pressure, heart rate, body temperature, and respiratory rate, etc., obtained from the patients. Prior to analysis, the collected data often undergoes cleaning and pre-processing steps. This involves identifying and rectifying data inconsistencies, handling missing values, removing outliers, and ensuring data quality and consistency.

The analysis of clinical measurements, including blood pressure, heart rate, body temperature, and respiratory rate, yields several valuable outputs. These outputs include statistical summaries to understand central tendencies and variabilities, evaluation of treatment responses to assess the drug's effects, identification of abnormalities or outliers that may require further attention, trend analysis to observe patterns over time, correlation analysis to explore relationships between measurements, detection of adverse events or side effects, subgroup analysis to assess variations among different population subsets, and data visualisations for easier interpretation and communication of results. These outputs collectively provide insights into the drug's impact on physiological parameters, its effectiveness, safety, and potential adverse effects within the clinical trial.

The data used for analysis, including the raw clinical measurements, would typically be stored using off-chain data storage solutions. These off-chain databases or systems are designed to securely manage and store the large volumes of data involved in clinical trials. However, the outputs obtained from the analysis, can be stored on the blockchain.

4.4.4 Blockchain Service Layer

The three main components in the blockchain service layer will be: blockchain storage, smart contracts, and consensus mechanisms.

1. *Blockchain Storage*

   Using blockchain storage for clinical drug testing analysis outputs provides several benefits. It ensures transparency, traceability, and immutability, preserving the integrity of results. Blockchain offers robust data security, privacy, and control over access rights.

It enhances auditability, facilitates collaborative research and reproducibility, and enables trustworthy data exchange. Compliance with regulatory requirements is simplified, promoting transparency and accountability in clinical research. Overall, leveraging blockchain storage improves the trustworthiness and reliability of analysis results, advancing transparency in clinical drug testing.

2. *Smart Contracts*

   Smart contracts in the blockchain service layer of clinical drug testing serve various purposes, including participant consent and data sharing, data access and permissions, data integrity and provenance, protocol compliance and governance, incentives and rewards, payment and compensation, and supply chain and drug traceability.

Some examples of smart contracts that can deployed in the clinical drug testing framework:

The first smart contract, named HealthcareData, facilitates the management of healthcare data on the blockchain. It allows for storing and retrieving patient information, managing medical states and critical conditions, and tracking devices associated with patients. The contract ensures transparency, security, and immutability of healthcare data by leveraging blockchain technology.

```
// SPDX-License-Identifier: MIT
pragma solidity ^0.8.0;

contract HealthcareData {

 uint public numberofpatients;
  struct Device {
    uint256 deviceId;
    string deviceType;
    uint256 patientId;
 }
  mapping(uint256 => Device[]) private patientDevices;
  address private patientRegistrationAddress;
  constructor(address _patientRegistrationAddress) {
    patientRegistrationAddress = _patientRegistrationAddress;
 }
```

```
function getPatientRegistrationAddress() public view returns
(address) {
return patientRegistrationAddress;
}

struct MedicalState {
   string condition;
   uint256 timestamp;
}

struct Patient {
   uint256 id;
   string name;
   uint256 age;
   string gender;
   MedicalState[] medicalStates;
   string doctorName;
}
mapping(uint256 => Patient) patients;

function addPatientData(
   uint256 _id,
   string memory _name,
   uint256 _age,
   string memory _gender,
   string memory _condition,
   string memory _doctorName

) public {
   Patient storage patient = patients[_id];
   require(patient.id == 0, "Patient with this ID already exists.");
   patient.id = _id;
   patient.name = _name;
   patient.age = _age;
   patient.gender = _gender;
   patient.doctorName = _doctorName;
   MedicalState memory medicalState;
   medicalState.condition = _condition;
   medicalState.timestamp = block.timestamp;
```

```
    patient.medicalStates.push(medicalState);
    numberofpatients++;
  }

 function addCriticalState(uint256 _id, string memory _condition)
public {
 require(_id > 0 && _id <= numberofpatients + 1, "Invalid patient
ID");

    Patient storage patient = patients[_id];

    MedicalState memory medicalState;
    medicalState.condition = _condition;
    medicalState.timestamp = block.timestamp;

  patient.medicalStates.push(medicalState);
 }
function getPatientData(uint256 _id)
  public
  view
   returns (
     uint256,
     string memory,
     uint256,
     string memory,
     MedicalState[] memory,
     string memory
)
{
Patient memory patient = patients[_id];

require(patient.id != 0, "Patient with this ID does not exist.");

return (
   patient.id,
   patient.name,
   patient.age,
```

```
    patient.gender,
    patient.medicalStates,
    patient.doctorName
);
}
function addDevice(
    string memory _deviceType,
    uint256 _patientId,
    uint256 _deviceId
) public {
  require(_patientId > 0, "Invalid patient ID.");
  patientDevices[_patientId].push(
    Device(_deviceId, _deviceType, _patientId)
);
}
function getPatientDeviceCount(uint256 _patientId)
    public
    view
    returns (uint256)
{
    return patientDevices[_patientId].length;
}
function getPatientDeviceDetails(uint256 _patientId)
    public
    view
    returns (uint256[] memory, string[] memory)
{
    uint256 deviceCount = patientDevices[_patientId].length;
    uint256[] memory deviceIds = new uint256[](deviceCount);
    string[] memory deviceTypes = new string[](deviceCount);

  for (uint256 i = 0; i < deviceCount; i++) {
    deviceIds[i] = patientDevices[_patientId][i].deviceId;
    deviceTypes[i] = patientDevices[_patientId][i].deviceType;
 }
 return (deviceIds,deviceTypes);
 }
}
```

The next contract called PatientRegistration, allows patients to register by associating their unique patient ID with an address on the blockchain. Once registered, the patient's ID and address are stored in a mapping. This contract ensures secure and transparent patient identification within decentralised healthcare systems or applications.

```
// SPDX-License-Identifier: MIT
pragma solidity ^0.8.0;
contract PatientRegistration
{
 mapping(uint256 => address) public patientIdToAddress;
 event PatientRegistered(uint256 patientId, address patientAddress);
 function registerPatient(uint256 _patientId) public {
  require(_patientId > 0, "Invalid patient ID.");
  require(
   patientIdToAddress[_patientId] == address(0),
   "Patient with this ID already exists."
 );
  patientIdToAddress[_patientId] = msg.sender;
  emit PatientRegistered(_patientId, msg.sender);
 }
}
```

One another contract can be AnalysisResultStorage, designed to store and retrieve analysis results associated with patient IDs and device addresses. It uses a mapping to store the analysis results, and provides functions to store and retrieve the results. The contract ensures that the analysis result is not empty before storing it. It offers a secure and transparent way to track and access analysis results within decentralised systems or applications.

```
// SPDX-License-Identifier: MIT
pragma solidity ^0.8.0;

contract AnalysisResultStorage {
 mapping(string => mapping(address => string)) private
analysisResultsMapping;
 event AnalysisResultStored(
    string patientID,
    address deviceAddress,
    string analysisResult
 );
 function storeAnalysisResult(
    string memory patientID,
    address deviceAddress,
    string memory result
 ) public {
   require(bytes(result).length > 0, "Analysis result cannot be empty");
   analysisResultsMapping[patientID][deviceAddress] = result;
   emit AnalysisResultStored(patientID, deviceAddress, result);
 }
 function getAnalysisResult(string memory patientID, address
deviceAddress)
   public
   view
   returns (string memory)
 {
   return analysisResultsMapping[patientID][deviceAddress];
 }
}
```

3. *Consensus Mechanism*

   In this example framework, we will consider using the PBFT consensus mechanism.

   PBFT is a consensus algorithm used in distributed systems. It ensures that nodes in the network can agree on the order of transactions and reach consensus, even in the presence of faulty or malicious nodes. PBFT uses a three-phase protocol (pre-prepare, prepare, and commit) to achieve consensus. It can tolerate up to a certain number of faulty nodes and incorporates message authentication to ensure

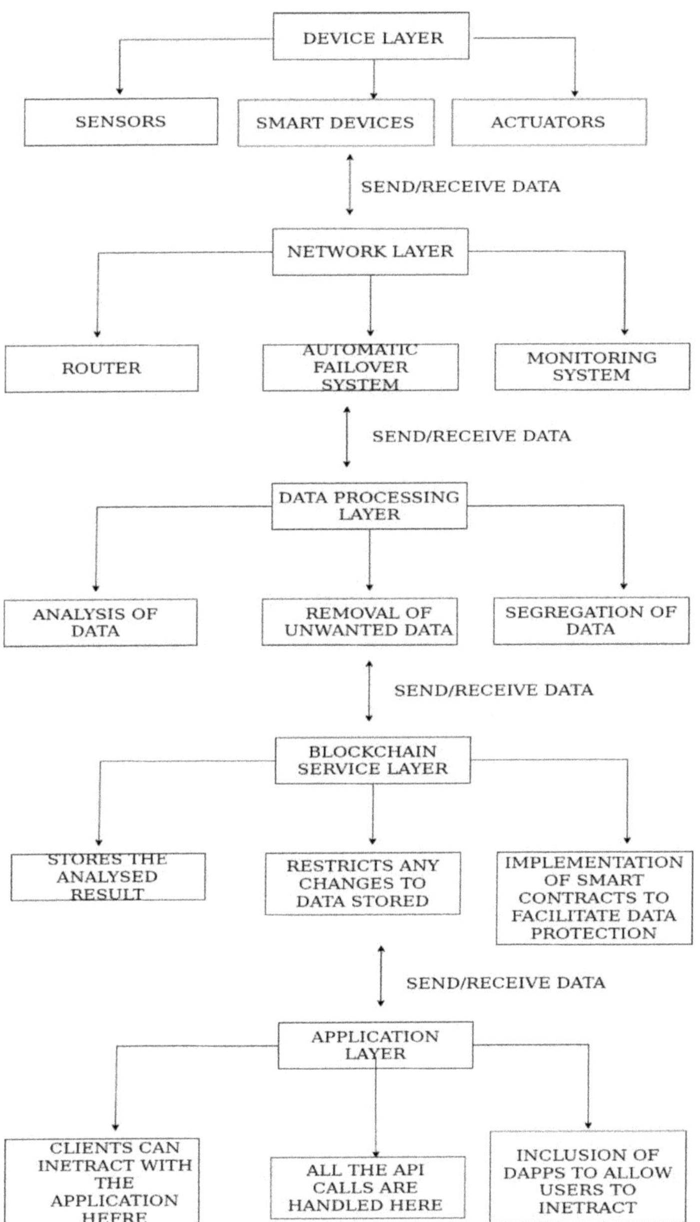

Fig. 3. IoT-blockchain model diagram for clinical drug testing.

the integrity of the communication. It provides a reliable and secure approach to consensus in Byzantine fault-tolerant systems.

If the PBFT consensus mechanism is employed in the blockchain service layer of clinical drug testing, it offers several advantages. The PBFT algorithm offers advantages in terms of quick transaction finalisation, energy efficiency, and low reward variation. It enables fast transaction processing, reduces energy consumption compared to other consensus mechanisms, and ensures a fair and balanced reward system for network participants. Additionally, PBFT enhances security through digital signatures and cryptographic techniques. This choice of consensus mechanism enhances the reliability, integrity, and efficiency of storing analysis results in a decentralised and secure manner.

4.4.5 Application Layer

The application layer in the clinical drug testing framework encompasses various components and functionalities. It involves a user interface for user interaction, account management for authentication and authorisation, data management for storing clinical trial data, consent management for participant consent handling, analysis and reporting tools for data analysis and report generation, collaboration and communication features, integration with external systems, and the utilisation of decentralised applications (DApps) to manage different aspects of the drug testing process. Data visualisation is also a part of the application layer, enabling users to analyse and interpret the stored data visually. Overall, the application layer acts as the interface between users and the underlying blockchain infrastructure, delivering the necessary tools and capabilities for efficient management and utilisation of clinical trial data.

5. Comparison of Blockchain Solutions vs. Classical Solutions

5.1 Challenges of Blockchain-based Solutions

5.1.1 Scalability

Scalability is a challenge in blockchain solutions. The decentralised nature and consensus mechanisms can limit transaction throughput and increase confirmation times as the number of participants and data size grows. Solutions like sharding, off-chain techniques, optimised

Table 2. Blockchain solutions vs. conventional approaches.

	Classical IoT Security Solutions	**Blockchain-based Solutions**
Security	Uses traditional security measures such as firewalls, encryption, authentication, and access control	Promotes the use of cryptographic techniques like hashing and consensus mechanisms to ensure data integrity and security
Single Point of Failure	There exists the issue of a weak node/hub which, if taken down, could destroy the whole network	No such issue exists here as each node is connected and if one node fails then also other nodes could function effectively
Data Privacy	Data privacy is a concern as sensitive information is stored in centralised databases	Provides enhanced data privacy by storing data in a decentralised manner and encrypting it
Immutability	Data can be altered or deleted by a central authority	Data is immutable and tamper-proof due to the use of cryptographic hashes and consensus mechanisms
Transparency	Limited transparency due to the centralised architecture	Offers transparency and accountability using a distributed ledger

consensus mechanisms, and layer 2 scaling are being explored to address scalability issues in blockchain for IoT.

5.1.2 Processing Power and Time

Performing big data analytics in the integrated IoT-blockchain system is challenging due to resource limitations and computing capabilities of IoT devices, as well as privacy protection mechanisms of blockchain. Uploading data to clouds for analysis can introduce latency and privacy concerns. Innovative approaches like edge computing and privacy-preserving analytics can help overcome these challenges and enable efficient analysis of IoT-generated data for valuable insights.

5.1.3 Storage

Storage is a significant aspect to consider in the integration of blockchain and IoT. While blockchain eliminates the need for a centralised server to store transactions and device IDs, the ledger itself needs to be stored on the participating nodes. As the number of nodes in the network increases and time progresses, the distributed ledger grows in size. This poses a challenge for IoT devices, which typically have limited computational resources and low storage capacity. Finding efficient ways to manage and store the growing blockchain ledger on resource-constrained IoT devices is crucial for the successful implementation of blockchain-based IoT solutions.

5.1.4 Legal and Compliance

Legal issues in blockchain solutions include regulatory compliance, data privacy conflicts, jurisdictional challenges, smart contract legality, intellectual property protection, AML and KYC compliance, governance and dispute resolution, and securities regulations. Addressing these issues requires the development of clear legal frameworks, collaboration between stakeholders, and international cooperation.

51.5 IoT Devices Mobility and Naming

Integrating blockchain and IoT poses a challenge in terms of IoT device mobility and naming. Blockchain networks are not designed for the dynamic nature of IoT devices, which are often mobile. Traditional approaches that rely on static IP addresses are not practical for IoT networks. Efficient naming and discovery mechanisms are needed to accommodate device mobility and enable seamless communication in a blockchain-based IoT system. Innovative solutions, such as decentralised naming systems and dynamic service discovery protocols, are being explored to address this challenge.

6. Conclusion

The application of Blockchain technology on the IoT presents a promising solution to the limitations and vulnerabilities that currently afflict existing IoT systems. While IoT has helped boost the production level and gained us huge profits, it is not a flawless system. Issues like lack of transparency, traceability, reliability, trustworthiness, and security pose significant risks to IoT networks, including single points of failure, data breaches, and attacks such as Man-in-the-Middle and DoS.

Thus, integrating blockchain technology with IoT presents a unique solution to counter these risks by providing a decentralised infrastructure that eliminates the above-mentioned risks. Furthermore, smart contracts can automate processes and eliminate human interference in the current architecture of IoT.

As IoT systems continue to gain momentum in various industries, the need for secure and trustworthy systems is paramount. Blockchain technology provides a viable solution for enhancing the security and reliability of IoT systems, opening doors to a more connected and secure future. However, it is significant to note that further research and development are required to improve the scalability and interoperability of Blockchain-based IoT systems to meet the demands of the ever-growing IoT ecosystem.

In conclusion, this chapter underscores the potential of blockchain technology in enhancing the security and reliability of IoT systems, representing a significant milestone toward creating a more secure and connected future.

References

[1] Shammar, E.A., Zahary, A.T. and Al-Shargabi, A.A. 2021. A survey of IoT and blockchain integration: security perspective. *IEEE Access*, 9: 156114–156150. Doi: 10.1109/ACCESS.2021.3129697.

[2] Ayub Khan, A., Laghari, A.A., Shaikh, Z.A., Dacko-Pikiewicz, Z. and Kot, S. 2022. Internet of Things (IoT) security with blockchain technology: a state-of-the-art review. *IEEE Access*, 10: 122679–122695. Doi: 10.1109/ACCESS.2022.3223370.

[3] Bahga, A. and Madisetti, V.K. 2016. Blockchain platform for industrial internet of things. *Journal of Software Engineering and Applications*, 9: 533–546. http://dx.doi.org/10.4236/jsea.2016.910036.

[4] Sadawi, A.A., Hassan, M.S. and Ndiaye, M. 2021. A survey on the integration of blockchain with IoT to enhance performance and eliminate challenges. *IEEE Access*, 9: 54478–54497. Doi: 10.1109/ACCESS.2021.3070555.

[5] Majid, A. 2023 Security and privacy concerns over IoT devices attacks in smart cities (2022). *Journal of Computer and Communications*, 11: 26–42. Doi: 10.4236/jcc.2023.111003.

[6] What is IoT Security (Internet of Things)?: Tools & Technologies. https://hackr.io/blog/what-is-iot-security-technologies.

[7] What is The Internet of Things or IoT? JFrog Connect. https://jfrog.com/connect/post/what-is-the-internet-of-things-or-iot-2021-real-examples/.

[8] Sadawi, A.A., Hassan, M.S. and Ndiaye, M. 2021. A survey on the integration of blockchain with IoT to enhance performance and eliminate challenges. *IEEE Access*, 9: 54478–54497. Doi: 10.1109/ACCESS.2021.3070555.

[9] What is Blockchain Technology? https://www.ibm.com/in-en/topics/what-is-blockchain.

[10] What is Blockchain Technology? https://en.wikipedia.org/wiki/Blockchain.

[11] Blockchain-Architecture. https://www.geeksforgeeks.org/blockchain-structure/.

[12] Idrees, Sheikh, Nowostawski, Mariusz, Jameel, Roshan and Mourya, Ashish. 2021. Security aspects of blockchain technology intended for industrial applications. *Electronics*, 10: 951. Doi: 10.3390/electronics10080951.

[13] Strebko, Julija and Romanovs, Andrejs. 2018. The Advantages and Disadvantages of the *Blockchain Technology*, 1–6. Doi: 10.1109/AIEEE.2018.8592253.

[14] Taherdoost, H. 2023. Smart contracts in blockchain technology: a critical review. *Information*, 14: 117. https://doi.org/10.3390/info14020117.

[15] What are public key and private keys? https://support.blockchain.com/hc/en-us/articles/4417082520724-What-are-public-and-private-keys-and-how-do-they-work-.

[16] Kouicem, Djamel Eddine, Bouabdallah, Abdelmadjid and Lakhlef, Hicham. 2018. Internet of Things Security: A top-down survey. *Computer Networks*, 141. Doi: 10.1016/j.comnet.2018.03.012.

[17] What are clinical trials and studies? https://www.nia.nih.gov/health/what-are-clinical-trials-and-studies.

[18] Sadawi, A.A., Hassan, M.S. and Ndiaye, M. 2021. A survey on the integration of blockchain with IoT to enhance performance and eliminate challenges. *IEEE Access*, 9: 54478–54497. Doi: 10.1109/ACCESS.2021.3070555.

[19] Atlam, Hany, Alenezi, Ahmed, Alassafi, Madini and Wills, Gary. 2018. Blockchain with Internet of Things: Benefits, challenges and future directions. *International Journal of Intelligent Systems and Applications*, 10. Doi: 10.5815/ijisa.2018.06.05.

[20] Qatawneh, Mohammad, Almobaideen, Wesam and Abualganam, Oraib. 2020. Challenges of blockchain technology in context internet of things: a survey. *International Journal of Computer Applications*, 175(16): 975–8887. Doi: 10.5120/ijca2020920660.

Chapter 6

A Comprehensive Assessment of Modern Blockchain-based Secured Health Monitoring Systems

*Swati Manekar** and *Umesh Bodkhe*

1. Introduction

Nowadays, the emphasis has switched from manual to electronic forms for the efficient preservation of clinical trials as well as personal health information (PHI). Data from patient EHRs (electronic health records) is typically gathered from places such as insurance companies, test results, prescription drug records, and medical sensors. Since the collected data is highly diverse, the industrial ecosystem progressively changed from Healthcare *1.0* to Healthcare *4.0* to preserve EHR. In Healthcare *1.0*, records were manually created and kept in files. Therefore, anyone with

Department of Computer Science and Engineering, Institute of Technology, Nirma University, Ahmedabad, Gujarat, India.

Email: umesh.bodkhe@nirmauni.ac.in

* Corresponding author: swati.manekar@nirmauni.ac.in

access could violate the patients' trust and privacy by accessing those manual records [2].

With the advent of Healthcare *2.0*, manual intervention was decreased and EHR data was kept electronically on centralised servers. The servers were open to network intrusions that may give hackers access to sensitive and vital medical data. Additionally, the servers were a single point of contact, necessitating the need for effective load balancing. EHR decentralisation using mobile apps for effective administration and retrieval of EHR was a trend in healthcare *3.0*. As a result, costs were cut, but the applications lacked the intelligence required for customised care. Decentralised intelligence was introduced by healthcare *4.0* to support decision analytics as well as real-time monitoring. Healthcare *4.0*, nevertheless, faces challenges with logistical management, record fragmentation, varied locations, sophisticated recommender models to promote personalization, as well as stakeholder trust issues.

Therefore, a decentralised trust amongst industry players is necessary to tackle the issues mentioned above in Healthcare *4.0*. Thus, BC (blockchain) in Healthcare *4.0* appears to be a revolution that can offer anonymous trust as a shared set of facts among all involved parties [3, 4]. A BC creates a distributed and chronological ledger to handle information sharing. On the global front, Onik [1] estimated that by 2030, there will be 1.6 million sufferers globally. The details of the technological revolution in the healthcare sector are shown in Fig. 1(a). The use of BC in Healthcare *4.0* can be observed in Fig. 1(b) [5].

A. Research Contribution

1) A comprehensive assessment of modern BC-based HMS are presented in detail.
2) A BC-based HMS is discussed in this paper.
3) Lastly, outstanding open issues, research possibilities, and future research avenues in BC-based HMS are presented.

B. Article Layout

The following provides a description of the paper's reminder parts. The associated work and survey paper's scope are presented in Section 2 of the manuscript. Section 3 presents background information and BC-based health monitoring methods. Section 4 outlines the research difficulties

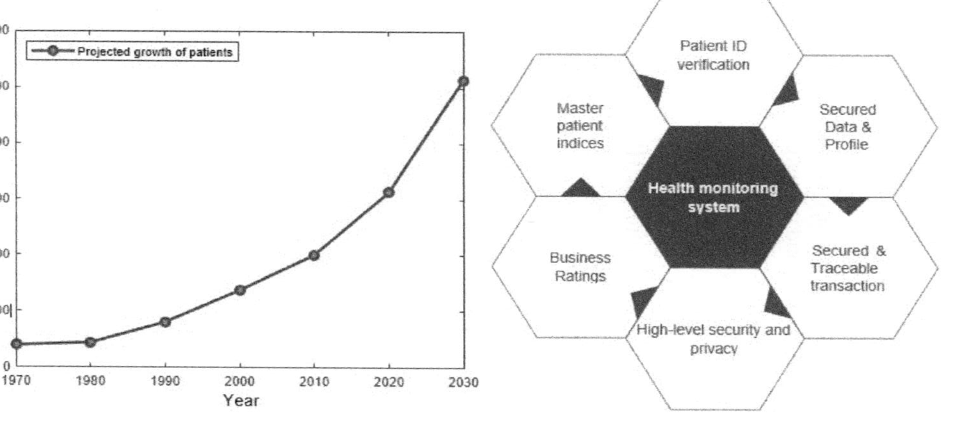

Fig. 1. (a) Projected worldwide growth of patients by 2030 [1], and (b) Pros of BC in Healthcare *4.0*.

and unresolved problems in smart healthcare. Finally, Section 5 eventually discusses the planned survey paper's concluding remarks.

2. Related Work

This section covers the most cutting-edge method for providing healthcare that is based on BC technology. Numerous studies have examined different security elements of MHR, PHR, and EHR data up to the present day. However, according to the literature we surveyed, a lot of these surveys mostly concentrated on secure and trustworthy approaches, strategies, and procedures. The majority of researches have taken into factors such patient encryption key, privacy-preserving methods, access control, exchange of medical data, distributed EHR, privacy, integrity, EHR security, verification, and validation. The comparative analysis of cutting-edge healthcare security standards useful for HMS is shown in Fig. 2. The authors of [6] proposed different BC-based IoT (internet of things) healthcare solutions as well as explored data sharing and storing. They also used a variety of security factors to assess how well the proposed approach performed.

3. Blockchain-Based Health Monitoring System

We discuss the advantages and integration of BC in HMS in this section.

A. Background

The primary function of Healthcare *4.0* is to offer patient monitoring and health diagnostics via an unguided media. From numerous IoT devices [7, 8], pertinent health-related information about various patients is gathered. Huge amounts of data were produced by heterogeneous devices, necessitating in-depth analysis and real-time monitoring. The patient data is extremely private and vulnerable to online threats [9, 10]. Additionally, we must protect it to ensure each patient's privacy. Medical stakeholders including doctors, hospitals, medicals, surgeons as well as patients share the analysed data. Therefore, safe and trustworthy communication among the numerous stakeholders in the healthcare *4.0* industry is essential for making important decisions [11]. Physician recommendations, planning for new hospital services, symptom analysis of various health-related difficulties, along with the system-wide improvement are among the main factors to consider.

Author	Year	Objective	1	2	3	4	5	6	7	8	Pros	Cons
Magyar et al. [1]	2017	To address privacy issues in Electronic Health Records (EHR) by integrating Blockchain (BC) with EHR database	✗	✓	✓	✓	✓	✓	✗	✗	Trust, security, speed, disintegration	Interchangeability, data integrity
Al-hadhrami et al. [2]	2017	To investigate the feasibility of Blockchain in healthcare	✓	✓	✓	✓	✓	✓	✗	✗	Easily organised data, consent management	Sybil attacks
Jiang et al. [3]	2018	To propose health-data-exchange, BC-based system	✓	✓	✗	✓	✗	✓	✓	✓	High-level privacy	Minimum throughput of the system
Theodouli et al. [4]	2018	To design a system for seamless healthcare data sharing	✗	✓	✓	✗	✗	✓	✓	✗	Automation and accountable workflow	Pseudo-anonymity, single point of failure
Li et al. [5]	2018	To explore the importance of healthcare data and its preventive mechanisms	✓	✗	✓	✓	✓	✓	✗	✓	Secured cryptographic solutions	Damage issue during data storage
Fan et al. [6]	2018	To propose efficient and tamper-resistant health data sharing with a Blockchain network	✗	✓	✓	✗	✓	✓	✓	✓	Privacy preservation	huge computation power.
Griggs et al. [7]	2018	To propose a Blockchain-based EHR system and smart contracts for remote patient monitoring	✗	✓	✓	✓	✓	✓	✓	✗	Real-time patient monitoring	Response time is minimum.
Sun et al. [8]	2018	To conduct an in-depth survey on attribute-based signature for healthcare using Blockchain	✗	✓	✓	✓	✓	✓	✓	✓	Non-repudiation	Less storage capacity
Saia et al. [9]	2019	To establish formal communication between entities and trackers through Blockchain	✓	✓	✓	✓	✓	✗	✗	✓	Secure communication	Enabling all devices currently to shift to IoE would be challenge
Hathaliya et al. [10]	2019	An algorithm for biometric-based authentication with effective communication and computational cost	✓	✓	✓	✓	✓	✓	✗	✓	Secure trusted communication in the Internet of Entities paradigm	Challenges in transitioning all devices to the Internet of Entities
McGhin et al. [11]	2019	Exploring the usability of Blockchain in various domains	✓	✓	✗	✓	✓	✓	✓	✓	Unexplored things many survey paper explained	Real world evaluation not done
Khezr et al. [12]	2019	Studying the deployment of Blockchain in healthcare	✓	✓	✓	✗	✓	✓	✓	✓	A very good comprehensive work	Detailing not done for some domains
Otoum et al. [13]	2019	Evaluation of Intrusion Detection Systems (IDS) for Wireless Sensor Network (WSN)-based critical monitoring infrastructure using Machine Learning (ML)/Deep Learning (DL)	✓	✓	✓	✗	✓	✓	✓	✓	Implemented multiple algorithm and shown comparative results	Very less results
Bhattacharya et al. [14]	2020	Integrating Blockchain with deep learning to enhance Electronic Health Record (EHR) security	✓	✓	✓	✗	✓	✓	✓	✗	Novel deep learning based recommender algorithm for patient illness	Experimental evaluation needs to be validated in real world scenarios
Manekar et al.	2022	A Comprehensive Assessment of Modern Blockchain-based Secured Health Monitoring Systems	✓	✓	✓	✓	✓	✓	✓	✓	A comprehensive survey	-
1: Architecture 2: Integrity of data 3: Data exchange 4: Access control mechanism 5: Distributed EHR 6: Encryption key for the patient 7: Simulation tool 8: Algorithm/Pseudo code												

Fig. 2. State-of-the-art BC-based approaches for healthcare monitoring system.

B. Blockchain-based Health Monitoring System

Numerous healthcare databases are currently kept on centralised client-server platforms. On such platforms, a single administrator or party controls all permissions and choices. Additionally, a single entity authenticates every stakeholder in the centralised healthcare system. Some of the frequent problems in centralised healthcare systems are server crashes, single points of failure, privacy concerns and security concerns. We can leverage BC-based solutions to overcome the drawbacks of these traditional centralised systems.

Data is stored and shared in a BC, which is an immutable linked list of blocks that is distributed, safe, and transparent [12, 13]. If one of the blocks in the chain is changed, the entire BC is disrupted since the cryptographic linkages are broken. As a result, it offers security and keeps transaction records in a way that can be checked. Because of this, there is no longer a need for trustworthy parties; even untrusted people or objects can interact safely. Blockchain technology provides a range of advantages as follows.

- *Process Integrity*: By utilising consensus mechanisms, blockchain ensures the integrity of processes. Transactions or records added to the blockchain are difficult to modify or tamper with, maintaining the trustworthiness of the system.
- *Traceability*: BC's transparent and immutable nature allows for comprehensive traceability. Each transaction recorded on the blockchain can be traced back to its source, facilitating transparency and accountability.
- *Disintermediation Security*: BC eliminates the need for intermediaries or centralised authorities, reducing the risk of manipulation or unauthorised access. The decentralised nature of BC enhances security and mitigates single points of failure.
- *Automation*: Smart contracts, powered by BC, enable automated execution of predefined actions once specified conditions are met. This automation streamlines processes, reduces manual intervention, and increases efficiency.
- *Immutability*: Once data is recorded on the BC, it cannot be altered or deleted without consensus from the network participants. This immutability ensures data integrity and reduces the possibility of fraud or data manipulation.
- *Trust*: BC's decentralised and transparent nature fosters trust among participants. By eliminating the need for intermediaries and providing a shared, tamper-resistant ledger, Bc enhances trust in transactions and data exchange.
- *Cost Efficiency*: Bc eliminates the need for third-party intermediaries, reducing costs associated with traditional processes. The decentralised nature of blockchain also minimises infrastructure and operational expenses.
- *Quicker Processing*: BC enables faster transaction processing and settlement, especially when compared to traditional systems that involve multiple intermediaries and manual processes. The removal of intermediaries and the automation of processes contribute to quicker transaction execution.

BC has undergone several evolutions. For instance, the first generation of BC (sometimes referred to as BC *1.0*) began with the bitcoin network in 2009. The main idea of this generation was cryptocurrency and payments. The hyperledger frameworks and smart contracts (i.e., Etheruem) were introduced in BC *2.0* [13, 14]. Only BC *3.0* made mobile apps useful for BC in an EHR for healthcare. The most recent era of BC innovation is BC *4.0*. It is used to create and run applications for Healthcare *4.0* in a useful state for business. Decentralised systems, like BC-based systems, can also eliminate some of the dangers related to the centralised control system. The decentralised BC stores the data and constructs the structural data storage, which increases the network's resilience in contrast to a centralised-based method. In conclusion, decentralised systems, like BC-based systems, can reduce the aforementioned hazards related to the centralised control system. To increase the network's impermeability, BC saves the data and creates the structural data storage.

The comparative analysis of cutting-edge healthcare security standards useful for Healthcare *4.0* is shown in Fig. 2. Various characteristics are used to differentiate the security standards. An access control system, privacy, cost reduction, security and many more factors are included. In the healthcare industry, exchange or sharing of the MHR, her, or PHR is essential for determining disease symptoms, potential appropriate treatments, medications, and for numerous healthcare related research projects. From the point of view of security, the current patient data access control measures are inadequate. Due to a lack of patient histories necessary for the right therapy, doctors occasionally do not have an acceptable or sufficient patient database. Even in cases of medical emergency, doctors occasionally do not have access to accurate patient histories due to unsafe patient data/record maintenance. These issues can be handled by employing BC to maintain PHR, MHR, and EHR in a secure manner.

Figure 3 presents a concept of the BC-based Healthcare monitoring system. Yue et al. [15] created the BC-based Healthcare Data Gateway (HDG) infrastructure, from which patients and doctors can safely share electronic health records. Figure 4 presents a procedure to store a health record in a BC network. Users who directly used the data include medical professionals, chemists, lab technicians, and the government. Reliable and trust-based storage layers guarantee the privacy and integrity of data. Metadata and schema management are efficiently maintained by the data

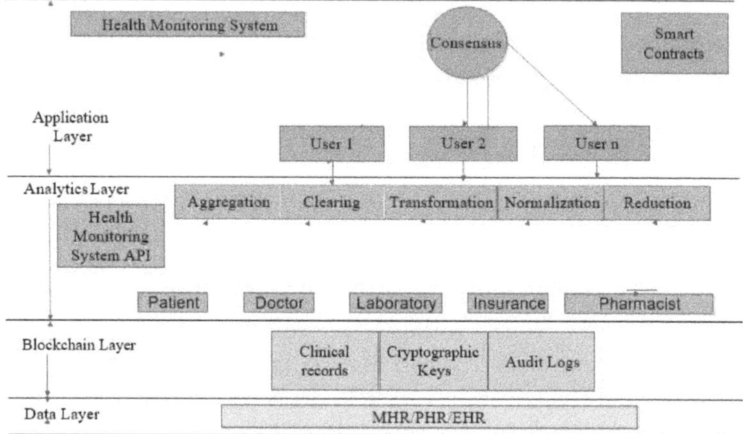

Fig. 3. BC-based healthcare monitoring system.

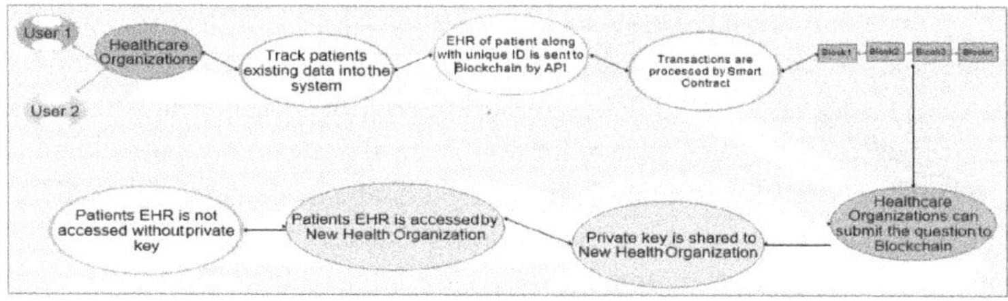

Fig. 4. Procedure to store a health record in BC network.

management layer. A BC-based MedRec model was created by Azaria et al. [16] to retrieve medical data using BC. The suggested approach safely distributes the medical data using BC. Shae et al. [17] designed BC-based precision medicine decentralised platform for clinical trial. It evaluates the data, protects user privacy, and preserves data integrity.

Zhang et al. [18] have proposed an attack-free BC-based social healthcare network built on the Pervasive Social Network (PSN) framework. This network utilises various protocols to ensure security and prevent attacks. PSN nodes exchange a medical database using the enhanced version of the protocol. The efficiency of transportation is constrained by the proposed model. In addition to addressing a framework for keeping and exchanging electronic medical information, specifically for cancer patients, Dubovitskaya et al. [19] concentrated on the BC-based healthcare system. By doing this, the system's overall cost and time for distributing EMR will be decreased. A completely BC-based system named MedShare was proposed and created by Xia et al. [20] to handle the massive volume of medical data on big data. Although the author concentrated on the topic of data privacy, other issues including scalability, data interoperability, and key management were not covered. Rifi et al. [21] concentrated on the benefit of employing BC-based technology for sharing medical records to improve performance while focusing on the issue of interoperability and scalability. A method was created and proposed by Liang et al. [22] to address the privacy and identity management issues associated with exchanging health information. A BC-based model for exchanging healthcare information was created by Jiang et al. [23], which in some ways addressed the aforementioned problems. In this suggested approach, the authors concentrated on health-related data and digital medical records and took into account the effectiveness of sharing and keeping medical records online in many different ways.

BC has been widely employed in studies geared towards the medical field during this technological revolution to manage, share and store health records securely. Theodouli et al. [24] concentrated on the application of this data for additional study and innovation in the industrial sector and the medical healthcare system. They [24] presented a layout of a system that ensures permission and administration of healthcare data with the aid of BC by examining the demands of the medical Healthcare system. The middleware, platform and BC network layers make up the three-layer system, with the first two layers being entirely cloud-based. When combined, these three layers improve the security and integrity of medical records. It was also used to check the data's exchange and compatibility within the system, which provided additional benefits like process automation, accountability, data integrity, and auditing. The infrastructure nodes connected wirelessly are mentioned in [25] as being able to communicate, but this connection is insecure without any safeguards. Saia et al. [25] suggested a BC-based strategy to protect this communication and establish trust between them. Additionally, this can obstruct communication with any unknown entity and affect the confidentiality. Bhattacharya et al. [26] place emphasis on edge computing in this situation. Compared to infrastructure based on the cloud, this decreases latency. They suggested a BC-based strategy for protecting mobile edge computing, which is performed on data generated by mobile nodes. Because it is based on BC and PoW was also provided for it, the framework provides a good level of trust and security.

The focus of authors of the citations [27] and [28] was on intrusion detection in wireless-based systems, such as smart cities. They investigated if this strategy could be used to target hospitals where patient data is transmitted wirelessly between a laboratory, a doctor, the front desk, and patients who are physically present in the building. Beyond simply authentication, authors in the citation [29] have conducted a survey on BC use in healthcare *4.0*. The main goal was to employ BC in healthcare specifically for fraud detection, medical incentives, scalability, and standardisation. The manuscript also emphasised on the utilisation of its applications, such as PSN, Omni PHR, MedRec, and Gem Network. The authors of [4, 30, 31] and [5] have investigated and validated many aspects of usefulness of BC. In contrast to Bodhke and Tanwar [4], who offered it in the tourism and hospitality industries, they have focused mostly on financial transactions that are occurring and protecting them with BC. They have also developed the equations for mining rewards and their distribution to miners that support BC infrastructure maintenance by giving their computing power.

4. Open Issues and Challenges

BC can use decentralised trust among healthcare stakeholders and add chronology to health records, as covered in the sections above. This increases user accountability across the Healthcare 4.0 ecosystem. As a result, the healthcare industry has used hypothetical scenarios to assess the advantages of BC-leveraged health systems, but widespread adoption and large-scale deployment are still a long way off. Lack of technical professionals, legal concerns with the use of cryptocurrencies, a lack of international conversion standards, untrained personnel for business promotion, and a general lack of trust among healthcare stakeholders regarding full adoption of [32] are the challenges facing the adoption of BC in Healthcare *4.0*. Because of this, despite many benefits, BC-envisioned healthcare systems are not implemented on the ground in reality. Figure 5 depicts a summary of the research obstacles and potential solutions presented by BC technology in Healthcare *4.0*.

Parameters	Challenges	Implications	BC-leveraged solutions
Master Patient Indices	Due to the redundant data and complexity of the EHR schema, the normalisation is inconsistent	Healthcare records are inconsistent, and each field's EHR schema is different	Single patient identifier is linked to effective hashed data structures with multiple keys, which produce effective search and access methods.
EHR Management	Third parties with unrestricted access to EHR records	Concerns about security and privacy since a malicious attacker could alter EHR records	After the patient has authorised the EHR, valid health blocks are added, and only authorised stakeholders with valid credentials are granted access.
Data Integrity	Complex linkages between the stored data, which makes the applications less responsive.	Difficulty in creation of efficient trained healthcare analytics and responsive queries	Timestamped ledger simplification, Chronological entries, fetch, as well as EHR record inquiry are persistent.
Traceability of Drugs	Counterfeiting drugs	Inadequate customer satisfaction	A timestamped, chronological ledger that is immutable and creates auditability for additional transactions.

Fig. 5. Research obstacles and potential solutions presented by BC technology in Healthcare *4.0*.

5. Conclusion

Academics and healthcare professionals have emphasised the need of IoT-based remote HMS for tracking the well-being of older people. Along with the enormous amounts of private patient-centric data created, HMS offers tremendous advantages. Due to outdated systems, insufficient cybersecurity, and close connectivity with access points, HMS are vulnerable to security threats, which poses a serious risk to patient data privacy. Numerous researchers proposed several security measures to overcome the HMS security concerns. If all security components (availability, confidentiality, and integrity) are not addressed, HMS will continue to pose a serious threat to forward secrecy, security, and privacy. A security framework utilising BC technology has been described in the literature as a means to provide high-level security in HMS. By leveraging BC technology, this framework has the potential to address various security attacks, thereby reducing the reliance on centralised authorities for performing different operations. Through the use of case studies, we compare and contrast the state-of-the-art security framework for the HMS provided by BC in this study. Along with the future directions, we additionally addressed about a number of unresolved problems and research difficulties.

References

[1] Onik, M.M.H., Aich, S., Yang, J., Kim, C.S. and Kim, H.C. 2019. Blockchain in healthcare: Challenges and solutions. pp. 197–226. *In*: *Big Data Analytics for Intelligent Healthcare Management*, Elsevier.

[2] Bodkhe, U., Tanwar, S., Parekh, K., Khanpara, P., Tyagi, S., Kumar, N. and Alazab, M. 2020. Blockchain for industry 4.0: A comprehensive review. *IEEE Access*, 8: 79764–79800.

[3] Bodkhe, U. and Tanwar, S. 2020, June. A taxonomy of secure data dissemination techniques for IoT environment. *IET Software*: 1–12.

[4] Bodkhe, U., Bhattacharya, P., Tanwar, S., Tyagi, S., mar, N. and Obaidat, M.S. 2019, Aug. Blohost: Blockchain enabled smart tourism and hospitality management. pp. 1–5. *In*: *2019 International Conference on Computer, Information and Telecommunication Systems (CITS)*.

[5] Bhattacharya, P., Tanwar, S., Bodkhe, U., Tyagi, S. and Kumar, N. 2019. Bindaas: Blockchain-based deep-learning as-a-service in healthcare 4.0 applications. *IEEE Transactions on Network Science and Engineering*, pp. 1–5.

[6] Nasab, S.S.F., Bahrepour, D. and Tabbakh, S.R.K. 2022. A review on secure data storage and data sharing technics in blockchain-based IoT healthcare systems. pp. 428–433. *In*: *2022 12th International Conference on Computer and Knowledge Engineering (ICCKE)*.

[7] Ladha, A., Bhattacharya, P., Chaubey, N. and Bodkhe, U. 2020. Iigpts: IoT-based framework for intelligent green public transportation system. pp. 183–195. *In*: Singh, P.K., Pawłowski, W., Kumar, N., Tanwar, S., Rodrigues, J.J.P.C. and Obaidat, M.S, (eds.). *Proceedings of First International Conference on Computing, Communications, and Cyber-Security (IC4S 2019)*, vol. 121. Springer International Publishing.

[8] Tanwar, S., Agarwal, B., Goyal, L. and Mittal, M. 2019. *Energy Conservation for IoT Devices: Concepts, Paradigms and Solutions*. Springer, 03.

[9] Bodkhe, U., Tanwar, S., Shah, P., Chaklasiya, J. and Vora, M. 2020. Markov model for password attack prevention. pp. 831–843. *In*: Singh, P.K., Pawłowski, W., Kumar, N., Tanwar, S., Rodrigues, J.J.P.C. and Obaidat, M.S. (eds.). *Proceedings of First International Conference on Computing, Communications, and Cyber-Security (IC4S 2019)*, vol. 121, Springer International Publishing.

[10] Tanwar, S. and Tyagi, N.K.S. 2019. *Multimedia Big Data Computing for IoT Applications: Concepts, Paradigms and Solutions*. Springer, 07.

[11] Tanwar, S. 2020. *Fog Computing for Healthcare 4.0 Environments*. Springer, 03.

[12] Bodkhe, U. and Tanwar, S. 2020, Feb. Secure data dissemination techniques for IoT applications: Research challenges and opportunities. *Software: Practice and Experience*, pp. 1–23.

[13] Bodkhe, U., Tanwar, S., Bhattacharya, P. and Kumar, N. 2020, June. Blockchain for precision irrigation: Opportunities and challenges. *Transactions on Emerging Telecommunications Technologies*, pp. 1–35.

[14] Tanwar, S., Singh, P., Kar, A., Singh, Y. and Kolekar, M. 2020. *Proceedings of ICRIC 2019: Recent Innovations in Computing*. Springer, 03.

[15] Yue, X., Wang, H., Jin, D., Li, M. and Jiang, W. 2016. Healthcare data gateways: Found healthcare intelligence on blockchain with novel privacy risk control. *Journal of Medical Systems*, 40(10): 218.

[16] Azaria, Ekblaw, A., Vieira, T. and Lippman, A. 2016, Aug. Medrec: Using blockchain for medical data access and permission management. pp. 25–30. *In*: *2016 2nd International Conference on Open and Big Data (OBD)*.

[17] Shae, Z. and Tsai, J.J.P. 2017, June. On the design of a blockchain platform for clinical trial and precision medicine. pp. 1972–1980. *In*: *2017 IEEE 37th International Conference on Distributed Computing Systems (ICDCS)*. June 2017.

[18] Zhang, J., Xue, N. and Huang, X. 2016. A secure system for pervasive social network-based healthcare. *IEEE Access*, 4: 9239–9250.

[19] Dubovitskaya, A., Xu, Z., Ryu, S., Schumacher, M. and Wang, F. 2017. *How Blockchain Could Empower eHealth: An Application for Radiation Oncology*, 9: 3–6.

[20] Xia, Q., Sifah, E., Omono Asamoah, K., Gao, J., Du, X. and Guizani, M. 2017. Medshare: Trustless medical data sharing among cloud service providers via blockchain. *IEEE Access*, PP(7): 1–11.

[21] Rifi, N., Rachkidi, E., Agoulmine, N. and Taher, N.C. 2017, Oct. Towards using blockchain technology for ehealth data access management. pp. 1–4. *In*: *2017 Fourth International Conference on Advances in Biomedical Engineering (ICABME)*.

[22] Liang, X., Zhao, J., Shetty, S., Liu, J. and Li, D. 2017, Oct. Integrating blockchain for data sharing and collaboration in mobile healthcare applications. pp. 1–5. *In*: *2017 IEEE 28th Annual International Symposium on Personal, Indoor, and Mobile Radio Communications (PIMRC)*.

[23] Jiang, S., Cao, J., Wu, H., Yang, Y., Ma, M. and He, J. 2018. Blochie: A blockchain-based platform for healthcare information exchange. pp. 49–56. *In*: *2018 IEEE International Conference on Smart Computing (SMART-COMP)*.

[24] Theodouli, A., Arakliotis, S., Moschou, K., Votis, K. and Tzovaras, D. 2018. On the design of a blockchain-based system to facilitate healthcare data sharing. pp. 1374–1379. *In*: *17th IEEE International Conference on Trust, Security and Privacy in Computing and Communications*, 08 2018.

[25] Saia, R., Carta, S., Recupero, D.R. and Fenu, G. 2019. Internet of entities (ioe): A blockchain-based distributed paradigm for data exchange between wireless-based devices. pp. 77–84. *In*: *Proceedings of the 8th International Conference on Sensor Networks - Volume 1: SENSORNETS*, INSTICC, SciTePress.

[26] Bhattacharya, P., Tanwar, S., Shah, R. and Ladha, A. 2020. Mobile edge computing-enabled blockchain framework: A survey. pp. 797–809. *In*: Singh, P.K., Kar, A.K., Singh, Y., Kolekar, M.H. and Tanwar, S. (eds.). *Proceedings of ICRIC 2019* (Cham), Springer International Publishing.

[27] Otoum, S., Kantarci, B. and Mouftah, H.T. 2019, June. On the feasibility of deep learning in sensor network intrusion detection. *IEEE Networking Letters*, 1: 68–71.

[28] Aloqaily, M., Otoum, S., Ridhawi, I.A. and Jararweh, Y. 2019. An intrusion detection system for connected vehicles in smart cities. *Ad Hoc Networks*, 90: 101842.

[29] McGhin, T., Choo, K.K.R., Liu, C.Z. and He, D. 2019. Blockchain in healthcare applications: Research challenges and opportunities. *Journal of Network and Computer Applications*, 135: 62–75.

[30] Kabra, N., Bhattacharya, P., Tanwar, S. and Tyagi, S. 2020. Mudrachain: Blockchain-based framework for automated cheque clearance in financial institutions. *Future Generation Computer Systems*, 102: 574–587.
[31] Srivastava, A., Bhattacharya, P., Singh, A., Mathur, A., Prakash, O. and Pradhan, R. 2018. A distributed credit transfer educational framework based on blockchain. pp. 54–59. *In*: *2018 Second International Conference on Advances in Computing, Control and Communication Technology (IAC3T),* Allahabad, India, IEEE.
[32] Tanwar, S., Tyagi, S. and Kumar, N. 2020. *Security and Privacy of Electronics Healthcare Records*. IET, UK, 03.

Chapter 7

Security of IoT Platforms

Current Challenges and Future Directions

*Rashi Sahay** and *Shanu Khare*

1. Introduction

In today's world the use for IoT devices has increased subsequently. With the human mind developing with each passing day, the need for new development that make the human life easy is also evolving. The need for automation of the day-to-day needs is also increasing. The moment we talk about automation, IoT devices and systems come to mind. To automate the daily needs of humans, IoT devices and systems have been introduced, such as **smart home automation systems, smart refrigerators, smart healthcare systems,** and so on. These systems work over the internet for various purposes such as receiving data over the network, storage of system data, and so on. To transfer or receive these requests and data over the internet, various network protocols are used. The most widely used protocol is the IP (Internet Protocol). As observed through implementation and use, this IP is not a reliable protocol and since all the connectivity and communication takes place

Computer Science and Engineering, Chandigarh University, Gharuan Mohali, India.
Email: shanukhare0@gmail.com
* Corresponding author: rashi.sahay787@gmail.com

over the internet, it can be said that the IoT systems and devices can be unreliable. This paper discusses the various security threats that the IoT systems tackle every day. The IoT systems are used in almost every household and therefore maintaining their security and integrity is of utmost importance. They face various threats that can be termed under hardware threats and software threats. This paper will discuss these threats and their solutions simultaneously.

A. History

In 1999, a computer scientist, Kevin Ashton, proposed the idea to induce radio frequency identification chips to track down products through huge supply chains for management and maintenance purposes. This idea was coined under the term '**Internet Of Things**'. The emergence of several IoT integrated systems thus started to grow and became an important necessity in day-to-day life. The very first smart refrigerator was launched by LG in 2000 and just after, the first iPhone was launched in 2007 and in time the number of smartly connected devices exceeded the number of people. The general public is surrounded by such applications and systems everywhere they go therefore the security of such platforms is a compulsion.

2. IoT Platform Architecture

IoT platforms are designed to handle a large amount of data generated by different types of devices. An IoT platform typically consists of three main component layers, namely, the device, the platform, and the application layers. The device layer constitutes of various IoT devices, such as sensors, actuators, and gateways. These devices collect and transmit data to the platform. The platform layer processes the data and provides various services, such as data analytic and visualisation, to the application layer which uses the services provided by the platform to develop IoT applications. This layered architecture provides a clear separation of concerns, which makes it easier to implement security measures at each layer.

A. Device Layer

This layer consists of a wide range of IoT devices, such as sensors, actuators, and gateways, which are responsible for collecting data from the physical world. These devices are connected to the platform through various communication protocols, such as Wi-Fi, Bluetooth, Zigbee, and cellular networks [11].

B. Platform Layer

This layer is responsible for processing the data collected by the devices in the device layer. The platform layer typically consists of several components, including data storage, data processing, data analytic, and data visualisation. Data storage components are responsible for storing the data collected from the devices, while data processing components are responsible for processing the data to generate insights and make decisions. Data analytic components use ML algorithms to analyse the data and provide insights. Data visualisation components provide a graphical representation of the data to make it easier to understand [11].

C. Application Layer

This layer is responsible for developing and deploying IoT applications. The application layer uses the services provided by the platform layer to create various applications, such as smart home systems, industrial automation systems, and healthcare monitoring systems. The applications developed in the application layer can be accessed through various interfaces, such as web applications, mobile applications, and APIs (application programming interface) [11].

3. IoT Security Threats

IoT security threats can be divided into two categories: physical threats and cyber threats. Physical threats include theft, tampering, and destruction of IoT devices, which can compromise the security of the entire platform. Cyber threats include unauthorised access, data breaches, and denial of service attacks, which can lead to the loss or theft of sensitive data. The increasing number of devices connected to the platform and the heterogeneity of these devices create a wide range of security challenges, making it difficult to detect and prevent security threats.

IoT security threats refer to a range of risks and vulnerabilities that can compromise various aspects of IoT devices such as confidentiality, integrity, and availability. These threats can arise from various sources, including malicious actors, software vulnerabilities, and operational errors. Some of the most common IoT security threats are described below.

A. Unauthorised Access

Unauthorised access is when an attacker gains illegal entry to an IoT device or system without proper authentication or authorisation. This can allow attackers to steal sensitive data, disrupt the operation of the device or system, or launch further attacks [3].

B. Malware

Malware refers to or can be termed as software that can damage or compromise IoT devices and systems. Malware can have many forms, including viruses, worms, and Trojan horses. Once installed on an IoT device, malware can be used to steal sensitive data, launch further attacks, or disrupt the operation of the device [3].

C. DDoS Attacks

Distributed Denial of Service (DDoS) attacks occur when a large number of IoT devices are hijacked and used to flood a target system with traffic, causing it to become overwhelmed and unavailable. These attacks can be used to disrupt critical infrastructure, such as power grids and communication networks [19].

D. Botnets

Botnets are networks of compromised IoT devices that are controlled by a central command and control server. Botnets can be used to launch a range of attacks, including DDoS attacks, spam campaigns, and malware distribution [19].

E. Data Breaches

Data breaches occur when sensitive data is stolen or leaked from an IoT device or system. This can include personal information, financial data, and intellectual property. Data breaches can have severe consequences, including identity theft, financial loss, and reputational damage [1].

F. Physical Attacks

Physical attacks occur when an attacker gains physical access to an IoT device and is able to manipulate or tamper with its components. Physical attacks can be used to extract sensitive data, modify the behaviour of the device, or render it inoperable [5].

G. Insider Threats

Insider threats occur when authorised users of an IoT system or device misuse their privileges to carry out malicious activities. This can include stealing sensitive data, sabotaging the system, or facilitating unauthorized access [1].

To address these IoT security threats, a range of security measures and best practices can be implemented. These include strong authentication and access controls, regular software updates and patches, network segmentation and isolation, encryption of sensitive data, and regular security audits and assessments. By adopting a proactive and comprehensive approach to IoT security, it is possible to mitigate the risks and vulnerabilities associated with IoT systems and devices.

4. IoT Platform Security Components

IoT platform security components are designed to address the security challenges faced by IoT platforms. These components include identity and access management, encryption and decryption, secure communication protocols, and intrusion detection and prevention systems. Identity and access management components ensure that only authorised users and devices have access to the platform. Encryption and decryption components protect sensitive data from unauthorised access. Secure communication protocols ensure that data transmitted between devices and the platform is secure. Intrusion detection and prevention systems monitor the platform for any suspicious activity and take appropriate action to prevent security threats [1].

IoT platform security components refer to the various tools, technologies, and processes that are used to secure the IoT platforms. These components are designed to address the unique security challenges associated with IoT platforms, such as the large number of devices and the diversity of communication protocols. The key IoT platform security components are described below.

A. Authentication and Access Controls

Authentication and access controls are used to ensure that only authorised devices and users are able to access the IoT platform. This includes using strong passwords, two-factor authentication, and digital certificates to verify the identity of devices and users. Authentication and access control are used to ensure that only authorised devices and users can access the IoT platform. Authentication verifies the identity of the device or user before granting access, while access control determines the level of access that is granted based on the identity of the device or user. Strong authentication and access control are essential for preventing unauthorised access to the IoT platform and protecting sensitive data. Common authentication methods used in IoT platforms include digital certificates, two-factor authentication, and biometric authentication.

B. Encryption

Encryption is used to protect data transmitted between devices and the IoT platform, as well as data stored on the platform. This includes using secure communication protocols, such as TLS (transport layer security), and implementing end-to-end encryption. Encryption is the process of encoding data so that it cannot be read by unauthorised parties. Encryption is an essential component of IoT platform security, as it ensures that sensitive data transmitted between devices and the platform, and data stored on the platform, is protected from interception and unauthorised access. Encryption can be implemented at different levels of the IoT platform, including the application layer, transport layer, and data layer. Common encryption methods used in IoT platforms include AES, RSA, and SSL/TLS.

C. Device Management

Device management is used to ensure that devices are properly configured, updated, and patched. This includes monitoring device activity, detecting anomalies, and remotely managing devices to prevent unauthorised access.

D. Data Privacy

Data privacy is used to ensure that sensitive data collected by IoT devices is protected from unauthorised access and use. This includes implementing privacy policies, data classification, and data anonymisation techniques. Data privacy is the protection of sensitive data collected by IoT devices from unauthorised access and use. Data privacy is a critical component of IoT platform security, as it ensures that sensitive data such as personal information, financial data, and health information is protected from hackers and other cyber threats. Data privacy can be achieved through data classification, data anonymisation, and the implementation of privacy policies [5].

E. Threat Detection and Response

Threat detection and response is used to detect and respond to security incidents on the IoT platform. This includes using security analytics, threat intelligence, and incident response plans to quickly identify and mitigate security threats. Threat detection and response is the process of identifying and responding to security threats on the IoT platform. Threat detection and response is essential for preventing cyber-attacks and mitigating the impact of security incidents. Threat detection and response can be achieved through the use of security analytics, threat intelligence, and incident response plans [5].

F. Network Segmentation and Isolation

Network segmentation and isolation is used to separate and protect different parts of the IoT platform from each other. Network segmentation and isolation are essential for preventing unauthorized access and the spread of security threats, and reducing the impact of security incidents, and can be achieved through the use of firewalls, VLANs, and DMZs.

G. Compliance and Auditing

Compliance and auditing is used to ensure that the IoT platform meets relevant regulatory and industry standards. This includes implementing security controls, conducting regular security assessments, and providing audit trails and logs for forensic analysis.

By implementing these IoT platform security components, it is possible to address the unique security challenges associated with IoT platforms and ensure the confidentiality, integrity, and availability of sensitive data. Additionally, it is important to adopt a risk-based approach to IoT platform security, which involves identifying and prioritising the most significant security risks and implementing appropriate security controls to mitigate them. In conclusion, IoT platform security components such as encryption, authentication and access control, data privacy, threat detection and response, and network segmentation and isolation are critical for ensuring the security and integrity of IoT platforms. By implementing these security components, organisations can protect sensitive data and critical infrastructure from cyber-attacks and demonstrate their commitment to security to customers and stakeholders.

5. IoT Security Standards

IoT security standards provide guidelines and best practices for implementing security measures in IoT platforms. These standards include ISO/IEC 27001, NIST Cybersecurity Framework, and the Industrial Internet Consortium (IIC) Security Framework. These standards provide a framework for implementing security measures in IoT platforms and ensure that these measures are consistent with industry best practices. IoT security standards refer to a set of guidelines and best practices for ensuring the security of IoT devices and systems. These standards are designed to provide a common framework for addressing the unique security challenges associated with IoT, such as the large number of devices, the diversity of communication protocols, and the need for interoperability. There are several IoT security standards that have been developed by various organisations and industry groups. Some of the most commonly used IoT security standards are described below.

A. IEC 62443

This is a series of standards developed by the International Electrotechnical Commission (IEC) that provides guidelines for securing industrial control systems (ICS), which are used in critical infrastructure such as power grids and manufacturing plants. IEC 62443 is a series of international standards for industrial automation and control systems (IACS) cyber security. The standards provide a comprehensive framework for protecting IACS networks and systems, including those used in IoT applications. The standards cover all aspects of cyber security, including risk assessment, security policies, access control, network segmentation, and incident response.

B. OWASP IoT Top 10

This is a list of the top 10 security risks associated with IoT devices and systems, developed by the Open Web Application Security Project (OWASP). It includes risks such as weak authentication, insecure communications, and lack of encryption. The OWASP IoT Top 10 is a list of the top 10 security vulnerabilities in IoT devices and applications. The list is intended to raise awareness of the most common security issues in IoT devices and to provide guidance for addressing these issues.

C. Zigbee Alliance

This is an industry group that develops and promotes the Zigbee wireless communication standard for IoT devices. The group also provides guidelines for securing Zigbee networks, including the use of encryption and key management. The Zigbee Alliance is a global organisation that develops and promotes open, global standards for the IoT. The Alliance has developed several security standards for IoT devices, including the Zigbee PRO Security, which provides secure communication between devices, and the Zigbee Trust Center, which manages device security.

D. Trusted Computing Group (TCG)

This is an industry group that develops and promotes trusted computing technologies, which are designed to ensure the integrity of data and systems. TCG provides guidelines for securing IoT devices and systems, including the use of hardware-based security and secure boot processes. It is an international organization that develops open standards for trusted computing. TCG has developed several security standards for IoT devices, including the Trusted Platform Module (TPM), which provides hardware-based security for IoT devices, and the Network Security Services (NSS), which provide secure communication between devices.

By following these IoT security standards, organisations can ensure that their IoT devices and systems are secure, and that they are able to protect sensitive data and critical infrastructure from cyber-attacks. Additionally, adherence to these standards can help organisations comply with relevant regulations and industry best practices, and demonstrate their commitment to security to customers and stakeholders. In conclusion, the development and adoption of IoT security standards are crucial for ensuring the security and integrity of IoT devices and applications. Standards such as ISO/IEC 27001, NIST Cybersecurity

Framework, IEC 62443, OWASP IoT Top 10, Zigbee Alliance, and Trusted Computing Group provide a common language and framework for managing cyber security risks associated with IoT devices and can help organisations to build secure and trustworthy IoT solutions.

6. Literature Review

Zhang, Zhi-Kai et al. [1], indulges in indicating the different security challenges that IoT devices face and also gives a detailed general review on these challenges and how the IoT devices encounter them. It is also diagnosed as to how the IoT devices will encounter these challenges in the future and what all challenges it can face in the coming time.

Abdullah et al. [2] explored the concept of blockchain and its security implementation techniques to help secure IoT devices and systems. They link blockchain methods to secure the IoT devices and how they can be implemented.

Frustaci et al. [3] SIoT which is a term used for integration of social networking platforms with IoT devices and systems. Since social media platforms need high security to maintain its user's privacy, and therefore the need for security of these IoT devices is of concern. They discuss the security paradigm of the three main layers of IoT systems namely perception, transportation, and application. Riahi et al. [4] proposed the IoT asks for a new paradigm of security, which will have to consider the security challenge from a holistic viewpoint encompassing the new actors and their interactions.

Hossain et al. [5] undertook a detailed analysis of IoT security issues and challenges, since they wanted to close this gap. They provide a thorough examination of the surfaces vulnerable to attacks, threat detection models, security concerns related to them, needs, forensics, and difficulties associated with IoT. To help researchers focus their efforts on the most pressing issues, they also offer a list of open topics in IoT security and privacy.

Gou et al. [6] began with the idea of the IoT, its characteristics, and systemic structure. They then provide an analysis of the security issue of IoT in the IoT system's perception, network, and application strata, propose a secure IoT construction, and provide corresponding secure strategies based on the IoT's current issues. And finally, the theoretical foundation will be made available to create a dependable IoT security solution.

Deogirikar et al. [7] analyse various IoT attacks that are occurring, classify them, discuss their defences, and identify the most significant IoT assaults. A state-of-the-art analysis of the different assaults, including their effectiveness and degree of damage in IoT, has been given and contrasted.

Ioannou et al. [8] suggest using the Support Vector Machine (SVM) learning anomaly detection model for the purpose of identifying anomalies within the IoT. Based on both good and bad local sensor activity, SVM generates its standard profile hyperplane. Real IoT network traffic combined with our customised network layer attacks is a key component of our research. This is in contrast to other works that use generic data sets to create supervised learning models. The suggested detection model achieves 81% accuracy while operating in an unknown topology and up to 100% accuracy when evaluated with unknown data drawn from the same network topology as it was trained.

Abomhara et al. [9] felt that although IoT cyber-attacks are nothing new, it is now more crucial than ever to take cyber defence seriously given how prevalent IoT is becoming in our daily lives and communities. As a result, there is a real need to secure IoT, which has necessitated a complete understanding of IoT infrastructure vulnerabilities and assaults. This article tries to categorise various threat types in addition to analysing and characterising attackers and incursions against IoT devices and services.

Mohindru et al. [10] furthermore addressed the IoT's security objectives and problems. To examine the effects of assaults on the network in more detail, certain IoT security attacks are analysed. There are many different forms of assaults in the IoT, some of which are simple to identify and stop, while others are quite challenging. The article aids in understanding the various facets of security concerns and security threats. The report also includes some future work that relates to the study areas.

Okul et al. [11] submitted that though the IoT does not yet have a comprehensive layer structure, it is stated that there are three generally agreed Object layer, Network layer, and Application layer, are layers: the Object layer, Network layer, and Application layer. Also, examples and analysis are used to illustrate the most prevalent IOT security epidemics, which include denial of service assaults, man-in-the-middle attacks, social engineering, data and identity theft, and botnet attacks. Lastly, these attacks outline the safeguards to take in the IOT's many tiers.

Sarigiannidis et al. [12] suggested a human-interactive visual-based anomaly detection system called VisIoT that is able to monitor and quickly identify a number of damaging security assaults, such as Sybil and wormhole attacks. VisIoT, which is built on a strict radial visualisation designed technique, may reveal attackers running one or more simultaneous attacks against IP-enabled WSNs (wireless sensor networks). Via a variety of simulated attack scenarios, the system's visual and anomaly detection effectiveness in uncovering complex security concerns is shown.

Xu et al. [13] begin by providing a succinct overview of IoT prospects and problems with a focus on security concerns. They then go through the possibility of IoT security methods based on hardware. The case studies that support the use of digital PPUFs (post purification filters) and stable PUFs (physical unclonable functions) for various IoT security protocols serve as the last examples in this paper.

Inayat et al. [14] discuss that both learning methods, that is, machine and deep learning are described and examined in connection to the detection of cyberattacks in IoT systems for learning-based approaches. To provide a clear picture of the many developments in this field, a thorough list of publications that have been made to date in the literature is integrated. The study also includes ideas for future research.

Bekri et al. [15] talk about and classify typical IoT attacks. Thus, they provide an assessment of the most recent countermeasures that have been developed to deal with the risky situations.

Singh [16] deliberate upon the blockchain concept together with pertinent elements that offer a thorough analysis of potential security threats. They also describe existing solutions that can be used as defences against such assaults. By distilling important ideas that can be used to create multiple blockchain systems and security monitoring tools and devices that address security flaws, this article also presents blockchain security enhancement options.

Qiu et al. [17] present two methods: A saliency map is then utilised to show how each packet property affects the detection results and the most important features. Then model extraction is used to duplicate the black-box model with a limited amount of training data.

Sandeep et al's. [18] study explores the security concerns and issues and provides a comprehensively defined security architecture as a secrecy of the client's privacy and safety, which may lead to a wider acceptance by the general public.

Liang et al. [19] believe that there will be significant property loss if the IoT system is hacked. This paper illustrates a denial of service (DoS) attack on the used IoT system/device. The Denial of Service (DOS) assault is launched using three different techniques using Kali Linux as the attack tool. Also provided is a comparison of the three DoS attack strategies.

Oh et al. [20] discussed IoT security needs, the main topic of this study. Initially, the authors analyse three fundamental characteristics to suggest IoT's fundamental security requirements (i.e., heterogeneity, resource constraint, dynamic environment).

7. Conclusions

The security of IoT platforms is a critical issue that needs to be addressed to ensure the privacy and security of sensitive data. This chapter has provided an updated overview of the security challenges faced by IoT platforms and the proposed solutions to address these challenges. The security of IoT platforms can be improved by implementing security measures at each layer of the platform, using IoT security components, and following IoT security standards. As the number of IoT devices continues to grow.

References

[1] Zhang, Z.K., Cho, M.C.Y., Wang, C.W., Hsu, C.W., Chen, C.K. and Shieh, S. 2014, November. IoT security: Ongoing challenges and research opportunities. pp. 230–234. *In*: *2014 IEEE 7th International Conference on Service-oriented Computing and Applications*. IEEE.

[2] Abdullah, A., Hamad, R., Abdulrahman, M., Moala, H. and Elkhediri, S. 2019, May. Cyber security: A review of internet of things (IoT) security issues, challenges, and techniques. pp. 1–6. *In*: *2019 2nd International Conference on Computer Applications Information Security (ICCAIS)*. IEEE.

[3] Frustaci, M., Pace, P., Aloi, G. and Fortino, G. 2017. Evaluating critical security issues of the IoT world: Present and future challenges. *IEEE Internet of Things Journal*, 5(4): 2483–2495.

[4] Riahi, A., Challal, Y., Natalizio, E., Chtourou, Z. and Bouabdallah, A. 2013, May. A systemic approach for IoT security. pp. 351–355. *In*: *2013 IEEE International Conference on Distributed Computing in Sensor Systems*. IEEE.

[5] Hossain, M.M., Fotouhi, M. and Hasan, R. 2015, June. Towards an analysis of security issues, challenges, and open problems in the internet of things. pp. 21–28. *In*: *2015 IEEE World Congress on Services*. IEEE.

[6] Gou, Q., Yan, L., Liu, Y. and Li, Y. 2013, August. Construction and strategies in IoT security system. pp. 1129–1132. *In*: *2013 IEEE International Conference on Green Computing and Communications and IEEE Internet of Things and IEEE Cyber, Physical and Social Computing*. IEEE.

[7] Deogirikar, J. and Vidhate, A. 2017, February. Security attacks in IoT: A survey. pp. 32–37. *In*: *2017 International Conference on I-SMAC (IoT in Social, Mobile, Analytics and Cloud) (I-SMAC)*. IEEE.

[8] Ioannou, C. and Vassiliou, V. 2019, May. Classifying security attacks in IoT networks using supervised learning. pp. 652–658. *In*: *2019 15th International Conference on Distributed Computing in Sensor* Systems *(DCOSS)*. IEEE.

[9] Abomhara, M. and Køien, G.M. 2015. Cyber security and the internet of things: vulnerabilities, threats, intruders and attacks. *Journal of Cyber Security and Mobility*: 65–88.

[10] Mohindru, V. and Garg, A. 2021. Security attacks in internet of things: A review. *Recent Innovations in Computing: Proceedings of ICRIC 2020*, pp. 679–693.

[11] Okul, S. and Aydın, M.A. 2017, October. Security attacks on IoT. pp. 1–5. *In*: *2017 International Conference on Computer Science and Engineering (UBMK)*. IEEE.

[12] Sarigiannidis, P., Karapistoli, E. and Economides, A.A. 2015, June. VisIoT: A threat visualisation tool for IoT systems security. pp. 2633–2638. *In*: *2015 IEEE International Conference on Communication Workshop (ICCW)*. IEEE.

[13] Xu, T., Wendt, J.B. and Potkonjak, M. 2014, November. Security of IoT systems: Design challenges and opportunities. pp. 417–423. *In*: *2014 IEEE/ACM International Conference on Computer-Aided Design (ICCAD)*. IEEE.

[14] Inayat, U., Zia, M.F., Mahmood, S., Khalid, H.M. and Benbouzid, M. 2022. Learning-based methods for cyberattacks detection in IoT systems: A survey on methods, analysis, and future prospects. *Electronics*, 11(9): 1502.

[15] Bekri, W., Layeb, T., Rihab, J.M.A.L. and Fourati, L.C. 2022, May. Intelligent IoT systems: Security issues, attacks, and countermeasures. pp. 231–236. *In*: *2022 International Wireless Communications and Mobile Computing (IWCMC)*. IEEE.

[16] Singh, S., Hosen, A.S. and Yoon, B. 2021. Blockchain security attacks, challenges, and solutions for the future distributed IoT network. *IEEE Access*, 9: 13938–13959.

[17] Qiu, H., Dong, T., Zhang, T., Lu, J., Memmi, G. and Qiu, M. 2020. Adversarial attacks against network intrusion detection in IoT systems. *IEEE Internet of Things Journal*, 8(13): 10327–10335.

[18] Sandeep, C.H. 2018. Security challenges and issues of the IoT system. *Indian Journal of Public Health Research Development*, 9(11).

[19] Liang, L., Zheng, K., Sheng, Q. and Huang, X. 2016, December. A denial of service attack method for an IoT system. pp. 360–364. *In*: *2016 8th International Conference on Information Technology in Medicine and Education (ITME)*. IEEE.

[20] Oh, S.R. and Kim, Y.G. 2017, February. Security requirements analysis for the IoT. pp. 1–6. *In*: *2017 International Conference on Platform Technology and Service (PlatCon)*. IEEE.

[21] Wehbi, K., Hong, L., Al-salah, T. and Bhutta, A.A. 2019, April. A survey on machine learning based detection on DDoS attacks for IoT systems. pp. 1–6. *In*: *2019 Southeast Conference*. IEEE.

[22] Abughazaleh, N., Bin, R. and Btish, M. 2020. DoS attacks in IoT systems and proposed solutions. *Int. J. Comput. Appl.*, 176(33): 16–19.

[23] Abosata, N., Al-Rubaye, S., Inalhan, G. and Emmanouilidis, C. 2021. Internet of things for system integrity: A comprehensive survey on security, attacks and countermeasures for industrial applications. *Sensors*, 21(11): 3654.
[24] Ghazal, T.M., Afifi, M.A.M. and Kalra, D. 2020. Security vulnerabilities, attacks, threats, and the proposed countermeasures for the Internet of Things applications. *Solid State Technology*, 63(1s).
[25] Bertino, E. 2016. March. Data security and privacy in the IoT. pp. 1–3. *In*: *EDBT*, Vol. 2016.

Chapter 8

Security Aspects of Blockchain Technology

Ramu Kuchipudi, T. Satyanarayana Murthy, Ramesh Babu Palamakula* and *R.M. Krishna Sureddi*

1. Introduction

In blockchain, data is kept in a distributed ledger. Blockchain technology provides integrity and availability allowing blockchain network participants to write, read, and verify transactions recorded in the distributed ledger. However, it does not allow deletion and modification of transactions and other information stored on its registry. The blockchain system is supported and secured by cryptographic principles and protocols, e.g., digital signatures, hash functions, etc. These principles ensure that transactions recorded in the ledger are integrity protected, verified for authenticity, and not rejected. Furthermore, as a distributed network, to allow all participants to agree on a unified record, blockchain technology also needs a consensus protocol, which is essentially a set of rules, for each participant, to reach a common agreement, unified opinion.

In a trustless environment, blockchain provides users with desirable characteristics of decentralisation, autonomy, integrity, immutability,

Associate Professor, Department of Information Technology, Chaitanya Bharathi Institute of Technology, Hyderabad, Telangana-500075.
* Corresponding author: kramupro@gmail.com

verification, fault tolerance, and has attracted great attention, of academia and industry in recent years. With these advanced features, blockchain technology has attracted great attention from academia and industry in the past few years. To help someone understand blockchain technology and blockchain security issues, especially for users who use blockchain to make transactions and for researchers who will develop blockchain technology and dealing with blockchain security issues, we have put in the effort and time to conduct a comprehensive investigation and analysis of blockchain technology and its security issues. First, identify keywords, namely blockchain, survey, consensus algorithm, smart contract, risk, and blockchain security to search for publications and information on the Internet. Second, blockchain-related investigative chapters are published in major security conferences and journals, e.g., USENIX Security Symposium, IEEE Security and Privacy Symposium, IEEE Transaction journals, etc. In this way, we interviewed as many chapters as possible to correct for biases in the study and results.

Distributed ledger or blockchain technologies are immutable digital ledger systems deployed in a distributed fashion (i.e., without a central repository) and often without a central authority. The technology became widely known in early 2008 when it was applied to give rise to the emergence of cryptocurrencies in which the transfer of digital money takes place in distributed systems. Various digital currency systems such as Bitcoin, Ethereum, Ripple, and Litecoin are just some examples of this technology. In fact, this technology is very useful and can be used for many different applications. Blockchain is a distributed digital ledger of cryptographically signed transactions grouped into blocks. Each block is cryptographically linked to the previous block after validation and is the subject of a consensus decision.

As new blocks are added, old blocks become harder to modify. New blocks are copied to all ledger replicas in the network and any conflicts are resolved automatically using established rules. Currently, there are mainly two types of blockchains: public blockchains and private blockchains. Public blockchains use computers connected to the public internet to validate transactions and group them into blocks to add to the ledger. Any computer connected to the Internet can join the party. Private blockchains, on the other hand, typically only allow known organisations to participate. Together, they form a private, members-only "business network". This distinction has important implications for where (potentially confidential) information circulating on the network is stored and who has access to it. Bitcoin is perhaps the most famous example of a public blockchain and it achieves consensus through 'mining'. In Bitcoin mining, network computers (or 'miners') attempt to solve a complex cryptographic problem to generate a proof of work.

The remainder of the chapter is organised as follows: Section 2 presents an overview. Section 3 describes security risks and attacks with Blockchain, and Section 4 discusses the real attacks and bugs on blockchain systems, while challenges and research trends are presented in Section 5. Finally, Section 6 concludes our work.

2. Overview of Blockchain History

In 1982, Chaum was the first known person to propose a blockchain-like protocol in his Ph. thesis [1–4]. In 1991, Haber and Stornetta cryptographically described a secure blockchain [5]. In 1993 Bayer et al. incorporated Merkle trees into the design [6]. In 1998, "bit gold"—a decentralised digital currency mechanism was devised by Szabo [7]. In 2008, Nakamoto introduced Bitcoin, a cryptocurrency with a pure peer-to-peer network [8]. Also in 2008, the term blockchain was first introduced as the distributed ledger behind Bitcoin transactions [9].

In 2013, Buterin proposed Ethereum in his white chapter [10–15]. In 2014, the development of Ethereum was crowdfunded, and on July 30, 2015, the Ethereum network went live. The emergence of Ethereum implies that blockchain 2.0 was born because unlike all other blockchain projects that focus on developing altcoins (other currencies similar to Bitcoin), Ethereum allows people to connect through distributed application without trusting its own blockchain. In other words, while Bitcoin was developed for a distributed ledger, Ethereum was developed for distributed data storage plus smart contracts, which are small computer programmes. Ethereum 2.0 upgrades the Ethereum network to increase network speed, scalability, efficiency, and security. The upgrade has 3 phases from 2020 to 2022.

In 2015, the Linux Foundation announced the Hyperledger project, which is an open-source software for blockchain. For the purpose of building an enterprise blockchain, the Hyperledger blockchain frameworks are different from Bitcoin and Ethereum. In Hyperledger, there are eight blockchain frameworks, including Hyperledger Besu, Hyperledger Fabric, Hyperledger Indy, Hyperledger Sawtooth, Hyperledger Burrow, Hyperledger Iroha, Hyperledger Grid, and

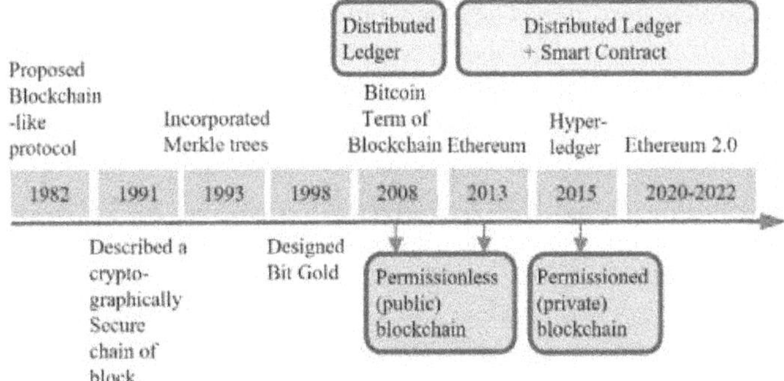

Fig. 1. History of blockchain.

Hyperledger Labs, five Hyperledger engines, including Hyperledger Avalon, Hyperledger Cactus, Heperledger Caliper, Hyperledger Cello, and Hyperledger Explorer, and four libraries, including Hyperledger Aries, Hyperledger Quilt, Hyperledger Transact, and Hyperledger URSA [16].

The history of the blockchain is summarised in the Fig. 1. Bitcoin and Ethereum are public blockchains because anyone can participate in their blockchain network, also known as permissionless blockchain. Various Hyperledger blockchain networks are private blockchains because participants must first be verified before joining the network, also known as Authorized Blockchain.

3. Security Risks and Attacks with Blockchain

Since the blockchain is decentralised without third-party involvement and trust must be ensured in the trustless infrastructure, the security on the blockchain itself is well worth the research. This section will focus on security risks on blockchain technology and examine actual attacks and bugs on blockchain systems. Top 10 web application security risks listed by OWASP Top 10 analysed and evaluated on blockchain technology [17–19]. The OWASP Top 10 is a widely known document on the top critical security risks in web applications and blockchain technology that face 9 of the top 10 [17–19]. Therefore, security on the blockchain is one of the key factors for success of blockchain business applications. A research team studied and analyzed vulnerabilities in the blockchain system from 2009 to May 2017 and listed 9 types of low-level blockchain security risks in [29]. Another research team has taken blockchain security to the next level. They point out that, like traditional computing, blockchain also faces denial of service (DoS) attacks, endpoint security, intentional abuse, code vulnerabilities, and data protection, but details of launch attacks vary [20–25].

In addition to DoS attacks, a number of research works have also presented BGP (Border Gate Protocol) hijacking attacks by manipulating routing advertisements, routing attacks by delaying block transmission or isolating parts of a blockchain network, overshadowing by isolating the victim from the network's view, EREBUS attacks by turning a malicious Transit Autonomous System (AS) into an intermediary network of Bitcoin nodes to infer decisions of the nodes such as stealthier attacks, DNS attacks, and remote channel attacks.

We categorise these attacks as cyber-attacks. Our chapter adds another type of risk from human negligence because humans are a weak point in any system. [20] lists six types of risks that an attacker can exploit to launch attacks.

Some other low-level security risks such as wallet security, attacks Sybil, private key security to emphasise its importance, and quick attacks, balance attacks, timejacking attacks, finney attacks, runaway attacks, and self-sustained attacks that we categorise Intentional abuse is also listed. In [21], it is clear that code vulnerabilities present the most risky surfaces on the blockchain. According to the code flaw, we split the code into the core software code on which blockchains 1.0 and 2.0 are built, and smart contracts that only exist in blockchain 2.0. According to the core software code, we emphasise the security of the wallet when some attacks hack the wallet. Blockchain is a very complex system and consists of distributed digital ledgers of cryptographically signed transactions grouped into blocks.

Blockchain has the following security features: (1) Blockchain technology relies on a ledger to keep track of all financial transactions. Usually, this kind of 'master' ledger would be a point of vulnerability. If the ledger is compromised, it can lead to system crashes. For example, if someone changes a record, they can steal an unlimited amount. Or, if they just read all transactions, then they may have access to sensitive personal information. In blockchain, the ledger is decentralised. This means that no computer or system has control of the ledger at any given time. It would take a highly sophisticated and coordinated attack on thousands of devices at once to get this kind of access to the master registry. (2) Another security principle is the chain itself.

The ledger exists as a long sequence of cryptographically encrypted sequential blocks. Each string represents another piece of the overall puzzle. Structurally, these records go back to the time of system launch. This means that anyone trying to modify a transaction must first modify all the transactions leading up to it and do so correctly. This makes the presumption spoofing process much more complicated.

In addition, it greatly increases the overall system security. (3) Unlike current payment systems, in the blockchain model there are hundreds or even thousands of distinct nodes. Each node has a complete copy of the digital ledger. They can work independently to verify transactions. If the nodes do not agree, the transaction is rolled back. This system keeps the registry neat. Moreover, due to its complex mechanisms, it is difficult to execute a fraudulent transaction, etc. The cryptographic keys as well as the two-key system used in blockchain exchanges are long, complex, and difficult to crack unless you have permission to view the keys.

Blockchain has a very complex and powerful structure. In this technology, though, there are the following security issues and challenges. Besides double spending, which is always possible in Bitcoin, the attack space includes a wide range of wallet attacks (i.e., client-side security), network attacks (such as DDoS, Sybil, and Eclipse) and exploit attacks (such as 50%, holdback and bribe).

Using a distributed ledger implies that data is shared among all the partners in the network. On the one hand, this can have a negative impact on security; on the other hand, it has a positive effect on availability with more nodes participating in the Blockchain, making it more powerful and flexible. Some of the traditional security challenges are: A. Key Management: Private keys are a direct means of authorising operations from an account, which, if accessed by an adversary, compromises any wallets or assets secured with those keys. Various potential private keys can be used to sign and encrypt messages in the distributed ledger. An attacker who obtains the encryption key from the dataset will be able to read the underlying data.

The private key is usually generated using a secure random function, which means reconstructing it is difficult, if not impossible. If a user loses the private key, any assets associated with that key will be lost.

4. Real Attacks and Bugs on Blockchain Systems

In this chapter, some actual attacks and bugs on the blockchain system in order to raise awareness of the need for security on the blockchain system will be discussed. Users use the exchange to transact on the blockchain, and on the blockchain, the private key is kept in a digital wallet. Therefore, exchanges and wallets are part of the blockchain system. Core Software Bug Appearing in August 2010, the CVE-2010-5139 vulnerability is the most notorious software bug in the Bitcoin network due to an integer overflow vulnerability in its protocol. Due to this error, an invalid transaction of 0.5 BTC replaced by 184 trillion BTC was added in a regular block and it took over 8 hours to fix this problem [25]. Also, when the Bitcoin version was upgraded from v0.7 to v0.8, an error occurred that caused a block processed in v0.8 to not be processed in v0.7 because the database used BerkeleyDB in v0.8 and LevelDB in v0.7. This bug causes different 6-hour blockchains to exist on nodes with v0.8 and nodes with v0.7 [29–30].

Cryptocurrency Exchange-Related Attacks: In 2011, attackers took several thousand BTC from Mt. Gox, a Bitcoin exchange based in Tokyo, due to network protocol loopholes and in March 2014, another 650,000 BTC in its online vault was stolen by hackers,causing Mt. Gox to file for bankruptcy due to a bug in the Bitcoin software that allowed users to change transaction IDs [31–32]. In December 2013, the anonymous Sheep Marketplace had to be shut down after it was announced that a website vendor exploited the vulnerability and stole 5,400 BTC. In August 2016, hackers stole 119,756 BTC from the third largest Bitcoin exchange, Bitfinex. In July 2020, hackers attacked Cashaa, a UK-based crypto exchange, and stole 336+ BTC. In August 2020, hackers attacked a cryptocurrency exchange platform in Europe, and stole $1.39 million. Wallet attack a user's wallet in a blockchain system stores their credentials and tracks digital assets associated with the user's address, credentials, and any other information other linked to their account. There have been attacks in the last 10 years.

Smart Contract Attacks and Bugs: A real life example of smart contract attacks is when a particular DAO (Decentralised Autonomous Organisation) smart contract is built on Ethereum for the fund. Crowd-based venture capitalist, a hacker exploited a weakness in its code and stole over $50 million worth of assets. **Cryptocurrency reported on June 17, 2016**: A hacker used a broken smart contract encryption to withdraw smart contract money. On June 19, 2016, Vitalik Buterin listed the following types of errors with Ethereum contracts: 1. variable/function name confusion, 2. public data that should not have been made public, 3. re-access (A calls B calls A), 4. send failure due to gas limit 2300, 5. arrays/loops and gas limits, and 6. subtle weaknesses in game theory. In January 2018, a hacker discovered an integer overflow bug with smart contracts using the Proof of Weak Hands (PoWH) coin and stole 888 ETH. In October 2018, an attacker launched a re-targeting attack on Spankchain smart contracts and withdrew 165.38 ETH. **Cyberattacks in August 2014**: A research team from the Dell SecureWorks Counter Threat Unit discovered that a BGP attacker was redirecting cryptocurrency miners' connections to a mining pool run by hackers who take control and profit from miners about $83,000 for four months or more. In September 2016, a DDoS (Distributed DoS) attack was discovered that attacked the Ethereum network such that an EXCODESIZE script was called approximately 50,000 times per block by the attacking transactions and thus slowing down substantial network. Endpoint Attacks Malware is part of endpoint attacks. According to the report, the malware has infected over a million computers, used by attackers to mine 26+ million crypto tokens. Another endpoint attack is cryptojacking, in which cryptocurrency is mined in a user's web browser when visiting a website.

Attackers hacked and injected cryptocurrency mining scripts into Pirate Bay, CBS Showtime in 2017 and the Indian government websites in 2018, and also won the Visitor Mining Award, accessed using the visitor's computer for mining. The attackers also injected cryptojacking code into third-party software (e.g., Google Tag Manager and WordPress in 2017 and Drupal in 2018 as well as advertising) (e.g., YouTube ads in 2018). Cryptojacking also occurred through 200,000 malware-infected MikroTik routers in 2018 and WiFi broke at a Starbucks coffee shop in Buenos Aires in 2017 to allow infected computers to mine for cryptocurrency. **IOTA Attacks**: In January 2019, a hacker carried out a phishing attack to collect users' private keys for six months and then steal $3.94 million worth of currency. At the same time, there was a DDoS attack on the IOTA network, so the IOTA developers were too busy to detect the hacker's stealing activity. In February 2020, to prevent attackers from stealing funds, the IOTA Foundation had to close the coordinator node for more than 12 days, the node responsible for confirming all transactions. The total mining value at current prices of BTC and ETH is over $40 billion. As a result, hackers have been, are and will continue to be encouraged to hack blockchain systems to reap great benefits.

5. Challenges and Research Trends

Several existing surveys have presented the future trends or prospects of blockchain technology. Blockchain testing, big data analysis, blockchain applications, smart contracts, centralisation prevention and artificial intelligence. A combined consensus mechanism, more efficient consensus, decryption of code, reliable computer execution against privacy leak risks, application enhancement, and an efficient mechanism for cleaning and detected data is presented. A standard testing mechanism, big data analysis, development and evaluation of smart contracts are provided in the reference [34]. Bug fixes in blockchain technology, more use cases and applications, and awareness of blockchain technology are mentioned in the reference [33]. In addition to these trends and ranges of validity, this chapter will highlight the research challenges and trends below.

6. Conclusion

This chapter conducts a more in-depth investigation of blockchain technology for the first time into an overview, consensus algorithms, smart contracts, and cryptography for blockchains. Public key cryptography, Zero-Knowledge Proof, and hash functions used in the blockchain have been described in detail about the integrity, authentication, non-repudiation, and remittances required in blockchain systems. This chapter then lists the complete blockchain applications. It goes on to showcase rich information and compare eight cryptocurrencies as the first blockchain application, the supply chain as the widespread use case, and the Dubai Smart Office as the service application, government first. In addition, the security of the blockchain itself is the focus of this chapter. It outlines comprehensive security risk categories based on the top 10 web application security risks, low-level risks, and high-level risks. This chapter then introduces security measures in the areas of security scanning, malware and error detection, software code security, maintaining confidentiality, and more. Specifically, he introduced and compared 11 smart contract bytecode vulnerability scanning tools.

Finally, research challenges and trends were presented for building more scalable and secure blockchain systems for mass deployment. We hope our efforts will help someone understand blockchain technology and blockchain security issues. Users who use the blockchain to make transactions pay more attention to the security of the blockchain itself.

We also hope that researchers will benefit so that they continue to study the development of blockchain technology and solve blockchain security issues.

References

[1] Saxena S., Gupta, U.K. and Renu, Dwivedi, V. 2022. Security issues and application of blockchain. *In*: *Lecture Notes in Electrical Engineering*, 768: 533–541, Springer Verlag.

[2] Alazzawi, M.Q. 2021. *Blockchain and Cyber Security*. Al-Mustaqbal University, pp. 1–20.

[3] Tan, M., Zheng, J., Zhang, J., Cheng, Y. and Huang, W. 2020. Research on security consensus algorithm based on blockchain. pp. 184–193. *In*: *Smart Computing and Communication: 5th International Conference, SmartCom 2020*.

[4] Guo, H. and Yu, X. 2022. A survey on blockchain technology and its security. *Blockchain Res. Appl.*, 3(2): 1–15.

[5] Islam, M.R., Rahman, M.M., Mahmud, M., Rahman, M.A., Mohamad, M.H.S, and Embong, A.H. 2021. A review on blockchain security issues and challenges. pp. 227–232. *In*: *IEEE 12th Control and System Graduate Research Colloquium (ICSGRC) 2021 Proceedings*.

[6] Government of India, Telecommunication Engineering Centre, Ministry of Communications and Information Technology, Department of Telecommunications. *Study Chapter on Security Aspects of Blockhain*, pp. 1–22.

[7] Gupta, S., Yadav, B. and Gupta, B. 2022. Security of IoT-based e-healthcare applications using blockchain. pp. 79–107. *In*: Maleh, Y., Tawalbeh, L., Motahhir, S. and Hafid, A.S. (eds.). *Advances in Blockchain Technology for Cyber Physical Systems*. Springer International Publishing, Cham.

[8] Li, X., Jiang, P., Chen, T., Luo, X. and Wen, Q. 2020 A survey on the security of blockchain systems. *Future Gener. Comput. Syst.*, 107: 841–853.

[9] Schuster, P., Theissen, E. and Uhrig-Homburg, M. 2020. Applications of blockchain technology in finance. *Schmalenbachs Zeitschrift fur Betriebswirtschaftliche Forsch*, 72(2): 125–147.

[10] Taylor, P.J., Dargahi, T., Dehghantanha, A., Parizi, R.M. and Choo, K.K.R. 2020. A systematic literature review of blockchain cyber security. *Digit. Commun. Netw.*, 6(2): 147–156.

[11] Phartyal, H. and Devi, S. 2022. Blockchain technology and its use cases. *Int. J. Res. Appl. Sci. Eng. Technol.*, 10(5): 4704–4713.

[12] Miller, D. 2018. Blockchain and the internet of things in the industrial sector. *IEEE Comput. Soc.*, 20(3): 15–18.

[13] Jose, D.T., Holme, J., Chakravorty, A. and Rong, C. 2022. Integrating big data and blockchain to manage energy smart grids: TOTEM framework. *Blockchain Res. Appl.*, 3(3): 100081.

[14] Liu, Y., Wang, Z. and Tian, S. 2019. Security against network attacks on web application system. *In*: *CNCERT Conference: Cyber Security,* 970: 145–152.

[15] Brent Lexi, Anton Jurisevic, Michael Kong, Eric Liu, Francois Gauthier, Vincent Gramoli, Ralph Holz and Bernhard Scholz 2018. *Vandal: A Scalable Security Analysis Framework for Smart Contracts*, Cornell University, pp. 1–28.

[16] Daneshgar, F., Ameri Sianaki, O. and Guruwacharya, P. 2019. Blockchain: A research framework for data security and privacy. *Advances in Intelligent Systems Computing,* 927: 966–974.

[17] Adeolu Seun, O. 2020. Blockchain technology for managing land titles in Nigeria. *Int. J. Adv. Trends Comput. Sci. Eng.*, 9(4): 5411–5417.

[18] Hussain, M.A., Abd Latiff, M.S., Madni, S.H.H., Raja Mohd Rasi, R.Z. and Othman, M.F.I. 2019. Concept of blockchain technology. *Int. J. Innov. Comput.*, 9(2).

[19] Banger, R., Mittal, R. and Khowal, R. 2019. A study on blockchain and cryptography. *J. Emerg. Technol. Innov. Res. (JETIR)*, 6(5): 27–31. ISSN: 2349-5162. Retrieved from.

[20] Banchhor, P., Sahu, D., Mishra, A. and Ahmed, M.B. 2021. A systematic review on blockchain security attacks, challenges, and issues. *Int. J. Eng. Res. Technol. (IJERT),* 10(04): 386–391. ISSN: 2278-081.

[21] Hasanaj, E. 2021. *Blockchain and Its Security Issues and Challenges.* Science Research Chapter, Middlesex University, London.

[22] Lage, O. and Saiz-Santos, M. 2021. Blockchain and the decentralisation of the cybersecurity industry. *Dyna. Ing. e Idustria. Collab.*, 93(3): 239–241. ISSN: 0012-7361.

[23] Krishnan, K.N., Jenu, R., Joseph, T. and Silpa, M.L. 2018. Blockchain-based security framework for IoT implementations. pp. 425–429. *In*: *Int. CET Conference on Control, Communication, and Computing (IC4).* IEEE. ISSN: 978-1-5386-4966-4.

[24] Zarrin, J., Wen Phang, H., Babu Saheer, L. and Zarrin, B. 2021. Blockchain for decentralization of internet: Prospects, trends, and challenges. *Cluster Comput.*, 24(4): 2841–2866.

[25] Shah, M., Shaikh, M., Mishra, V. and Tuscano, G. 2020. Decentralized cloud storage using blockchain. *In*: *2020 4th International Conference on Trends in Electronics and Informatics (ICOEI).*

[26] Wilczyński, A. and Widłak, A. 2019. Blockchain networks-security aspects and consensus models. *J. Telecommun. Inf. Technol.*, 2: 46–52.

[27] Zhai, S., Yang, Y., Li, J., Qiu, C. and Zhao, J. 2019. Research on the application of cryptography on the blockchain. *J. Phys. Conf. Ser.*, 1168(3): 1–8. 032077.

[28] Saini, V. 2019. Blockchain in supply chain: Journey from disruptive to sustainable. *JMCMS,* 14(2).

[29] Sachdeva, V. and Gupta, S. 2018. Basic NOSQL injection analysis and detection on MongoDB. pp. 1–5. *In*: *2018 International Conference on Advanced Computation and Telecommunication. (ICACAT).*

[30] Yadav, M. and Gupta, S. 2020. Hybrid metaheuristic VM load balancing optimization approach. *J. Inf. Optim. Sci.*, 41(2): 577–586.

[31] Kalra, S., Gupta, S. and Prasad, J.S. 2020. Predicting trends of stock market using SVM: A big data analytics approach. pp. 38–48. *In*: Batra, U., Roy, N.R. and Panda, B. (eds.). *Data Science and Analytics*, 1229. Springer Singapore, Singapore.

[32] Gupta, S. and Gupta, B. 2019. Performance modelling and evaluation of transportation systems using analytical recursive decomposition algorithm for cyclone mitigation. *J. Inf. Optim. Sci.*, 40(5): 1131–1141.

[33] Gupta, S. and Gupta, B. 2020. Securing honey supply chain through blockchain. pp. 321–335. *In*: Sharma, S.K., Bhushan, B. and Debnath, N.C. (eds.). *IoT Security Paradigms and Applications*, 1st Edn. CRC Press.

[34] Yadav B. and Gupta, S. 2022. Healthcare transformation traditional to telemedicine: a portrayal. pp. 3–14. *In*: Singh, P.K., Kolekar, M.H., Tanwar, S., Wierzchoń, S.T. and Bhatnagar, R.K. (eds.). *Emerging Technologies for Computing, Communication and Smart Cities*, 875. Springer Nature Singapore, Singapore.

Index

4-layered architecture 69
5G 14

A

Access control 102, 103, 105
Authentication 70–72, 79, 92, 94, 95
Availability 72, 81

B

Behavior analytics 52
Bitcoin 129–134
Blockchain 50, 52–58, 60–64, 99, 100, 102–104
Blockchain security 129, 131, 136, 137
Blockchain technology 6, 9, 10, 12, 13, 16, 17, 23, 52–57, 62–64, 67, 75, 79, 80, 87, 96, 97, 104, 128, 129, 131, 132, 136, 137
Botnets 116, 123

C

CIA Triad 2
Computer vision 26, 27, 30–47
Consensus 67, 68, 74, 75, 79–81, 86, 92, 94, 95
Consortium blockchain 74
Convolution neural network 31
Cryptocurrency 130, 134, 135
Cyber-attacks 5, 6, 8–10
Cybersecurity 50–60, 62–64, 120, 122

D

Data integrity 104, 106, 107
Decentralization 79
Deep learning 30, 31, 36, 38, 40, 46, 47
Digital signature algorithm 72
Distributed denial of service 5, 71
Distributed ledger 1
Dynamic Quorum-Based Access Control 72

E

Encryption 102, 103
Ethereum 129–131, 134, 135

F

False data injection 5, 7–10

H

Healthcare 99–103, 105–109
Healthcare systems 103, 106, 107
Hyperledger 130, 131

I

Internet of Things (IoT) 6, 13, 50–58, 60–64, 66, 68, 102, 114
IoT systems 114, 117, 122, 124, 125

M

Malware 116
Merkle Tree 2, 3

O

Open CV 27

P

Privacy 51, 53, 54–57, 59–64, 66–68, 70–72, 74, 81, 84, 86, 95, 96, 100, 102, 103, 105–107, 109

S

Security 102–109
Security and privacy 122
Security threats 114–117, 119, 123, 124
Security vulnerabilities 66
Sensors 114, 123, 124
Smart grids 5–10
Smart homes 51, 54, 113, 115
Smart parking 30
Software defined network 12, 13

T

Threat intelligence 52, 54–57, 60
Threats 135
Traceability 67, 70, 76, 80, 84, 86, 87, 96
Transparency 67, 71, 77, 79, 84, 86, 87, 95, 96

V

Vehicular ad-hoc networks 14

For Product Safety Concerns and Information please contact our EU representative GPSR@taylorandfrancis.com
Taylor & Francis Verlag GmbH, Kaufingerstraße 24, 80331 München, Germany

www.ingramcontent.com/pod-product-compliance
Ingram Content Group UK Ltd.
Pitfield, Milton Keynes, MK11 3LW, UK
UKHW021823240425
457818UK00006B/54